LONDON MATHEMATICAL SOCIETY LECTURE NOTE SERIES

1. General cohomology theory and K-theory, P.HILTON
4. Algebraic topology, J.F.ADAMS
5. Commutative algebra, J.T.KNIGHT
8. Integration and harmonic analysis on compact groups, R.E.EDWARDS
9. Elliptic functions and elliptic curves, P.DU VAL
10. Numerical ranges II, F.F.BONSALL & J.DUNCAN
11. New developments in topology, G.SEGAL (ed.)
12. Symposium on complex analysis, Canterbury, 1973, J.CLUNIE & W.K.HAYMAN (eds.)
13. Combinatorics: Proceedings of the British Combinatorial Conference 1973, T.P.McDONOUGH & V.C.MAVRON (eds.)
15. An introduction to topological groups, P.J.HIGGINS
16. Topics in finite groups, T.M.GAGEN
17. Differential germs and catastrophes, Th.BROCKER & L.LANDER
18. A geometric approach to homology theory, S.BUONCRISTIANO, C.P. ROURKE & B.J.SANDERSON
20. Sheaf theory, B.R.TENNISON
21. Automatic continuity of linear operators, A.M.SINCLAIR
23. Parallelisms of complete designs, P.J.CAMERON
24. The topology of Stiefel manifolds, I.M.JAMES
25. Lie groups and compact groups, J.F.PRICE
26. Transformation groups: Proceedings of the conference in the University of Newcastle-upon-Tyne, August 1976, C.KOSNIOWSKI
27. Skew field constructions, P.M.COHN
28. Brownian motion, Hardy spaces and bounded mean oscillations, K.E.PETERSEN
29. Pontryagin duality and the structure of locally compact Abelian groups, S.A.MORRIS
30. Interaction models, N.L.BIGGS
31. Continuous crossed products and type III von Neumann algebras, A.VAN DAELE
32. Uniform algebras and Jensen measures, T.W.GAMELIN
33. Permutation groups and combinatorial structures, N.L.BIGGS & A.T.WHITE
34. Representation theory of Lie groups, M.F. ATIYAH *et al.*
35. Trace ideals and their applications, B.SIMON
36. Homological group theory, C.T.C.WALL (ed.)
37. Partially ordered rings and semi-algebraic geometry, G.W.BRUMFIEL
38. Surveys in combinatorics, B.BOLLOBAS (ed.)
39. Affine sets and affine groups, D.G.NORTHCOTT
40. Introduction to Hp spaces, P.J.KOOSIS
41. Theory and applications of Hopf bifurcation, B.D.HASSARD, N.D.KAZARINOFF & Y-H.WAN
42. Topics in the theory of group presentations, D.L.JOHNSON
43. Graphs, codes and designs, P.J.CAMERON & J.H.VAN LINT
44. Z/2-homotopy theory, M.C.CRABB
45. Recursion theory: its generalisations and applications, F.R.DRAKE & S.S.WAINER (eds.)
46. p-adic analysis: a short course on recent work, N.KOBLITZ
47. Coding the Universe, A.BELLER, R.JENSEN & P.WELCH
48. Low-dimensional topology, R.BROWN & T.L.THICKSTUN (eds.)

49. Finite geometries and designs, P.CAMERON, J.W.P.HIRSCHFELD & D.R.HUGHES (eds.)
50. Commutator calculus and groups of homotopy classes, H.J.BAUES
51. Synthetic differential geometry, A.KOCK
52. Combinatorics, H.N.V.TEMPERLEY (ed.)
53. Singularity theory, V.I.ARNOLD
54. Markov processes and related problems of analysis, E.B.DYNKIN
55. Ordered permutation groups, A.M.W.GLASS
56. Journées arithmétiques 1980, J.V.ARMITAGE (ed.)
57. Techniques of geometric topology, R.A.FENN
58. Singularities of smooth functions and maps, J.MARTINET
59. Applicable differential geometry, M.CRAMPIN & F.A.E.PIRANI
60. Integrable systems, S.P.NOVIKOV et al.
61. The core model, A.DODD
62. Economics for mathematicians, J.W.S.CASSELS
63. Continuous semigroups in Banach algebras, A.M.SINCLAIR
64. Basic concepts of enriched category theory, G.M.KELLY
65. Several complex variables and complex manifolds I, M.J.FIELD
66. Several complex variables and complex manifolds II, M.J.FIELD
67. Classification problems in ergodic theory, W.PARRY & S.TUNCEL
68. Complex algebraic surfaces, A.BEAUVILLE
69. Representation theory, I.M.GELFAND et al.
70. Stochastic differential equations on manifolds, K.D.ELWORTHY
71. Groups - St Andrews 1981, C.M.CAMPBELL & E.F.ROBERTSON (eds.)
72. Commutative algebra: Durham 1981, R.Y.SHARP (ed.)
73. Riemann surfaces: a view towards several complex variables, A.T.HUCKLEBERRY
74. Symmetric designs: an algebraic approach, E.S.LANDER
75. New geometric splittings of classical knots (algebraic knots), L.SIEBENMANN & F.BONAHON
76. Linear differential operators, H.O.CORDES
77. Isolated singular points on complete intersections, E.J.N.LOOIJENGA
78. A primer on Riemann surfaces, A.F.BEARDON
79. Probability, statistics and analysis, J.F.C.KINGMAN & G.E.H.REUTER (eds.)
80. Introduction to the representation theory of compact and locally compact groups, A.ROBERT
81. Skew fields, P.K.DRAXL
82. Surveys in combinatorics: Invited papers for the ninth British Combinatorial Conference 1983, E.K.LLOYD (ed.)
83. Homogeneous structures on Riemannian manifolds, F.TRICERRI & L.VANHECKE
84. Finite group algebras and their modules, P.LANDROCK
85. Solitons, P.G.DRAZIN
86. Topological topics, I.M.JAMES (ed.)
87. Surveys in set theory, A.R.D.MATHIAS (ed.)
88. FPF ring theory, C.FAITH & S.PAGE
89. An F-space sampler, N.J.KALTON, N.T.PECK & J.W.ROBERTS
90. Polytopes and symmetry, S.A.ROBERTSON
91. Classgroups of group rings, M.J.TAYLOR
92. Simple artinian rings, A.H.SCHOFIELD
93. General and algebraic topology, I.M.JAMES & E.H.KRONHEIMER
94. Representations of general linear groups, G.D.JAMES
95. Low dimensional topology 1982: Proceedings of the Sussex Conference, 2-6 August 1982, R.A.FENN (ed.)
96. Diophantine equations over function fields, R.C.MASON

London Mathematical Society Lecture Note Series: 96

Diophantine Equations over Function Fields

R.C. MASON

Fellow of Gonville and Caius College, Cambridge

CAMBRIDGE UNIVERSITY PRESS
Cambridge
London New York New Rochelle
Melbourne Sydney

CAMBRIDGE UNIVERSITY PRESS
Cambridge, New York, Melbourne, Madrid, Cape Town,
Singapore, São Paulo, Delhi, Tokyo, Mexico City

Cambridge University Press
The Edinburgh Building, Cambridge CB2 8RU, UK

Published in the United States of America by Cambridge University Press, New York

www.cambridge.org
Information on this title: www.cambridge.org/9780521269834

First published 1984

A catalogue record for this publication is available from the British Library

Library of Congress catalogue card number: 84-1900

ISBN 978-0-521-26983-4 Paperback

Frau Minne will :

es werde Nacht,

dass hell sie dorten leuchte,

wo sie dein Licht verscheuchte.

Die Leuchte,

und wär's meines Lebens Licht,-

lachend

sie zu löschen zag ich nicht!

CONTENTS

PREFACE

At the International Congress of Mathematicians held in Paris in 1900, David Hilbert proposed a list of 23 problems intended to engage the attention of researchers in mathematics. The tenth problem requested a universal algorithm for deciding whether any Diophantine equation is soluble in integers. Although as proved by Matijasevic in 1970, no such general algorithm actually exists, this century has generated a substantial body of knowledge on the more limited, but tractable, problems associated with equations in two variables only. In particular, the works of Thue, Siegel and Baker, which are discussed within in some detail, have been particularly fruitful, and mention should also be made on a most recent outstanding result, namely Faltings' establishment of the Mordell conjecture.

The aim of this book is to provide a connected and self-contained account of researches on similar problems appertaining to equations over function fields. The purpose throughout is the production of complete algorithms for the effective resolution of the families of equations under consideration. The algorithms obtained turn out to be much simpler and more efficient than those provided by the analysis for number fields. This is a reflection of the strength of the fundamental inequality which forms the crux of the approach, and which may also be regarded as a loose analogue of Baker's celebrated lower bounds for linear forms in logarithms. The particular families of equations treated are those of the Thue, hyperelliptic, genus zero and genus one types. Several worked examples are given illustrating the general methods. A separate discussion is conducted on the extra problems arising for fields of positive characteristic, which problems are resolved. The final chapter is devoted to a novel application of the fundamental inequality and, hopefully, points towards future research in the area.

The first three chapters have already appeared as separate papers in slightly transmuted form; the remaining chapters have not been published before. Parts of the book formed a Smith's Prize Essay in 1981, a majority a Research Fellowship Dissertation for St. John's College in 1982, and the greatest part a Ph.D. Thesis in the University of Cambridge in 1983. I have also given several lectures on the subject in England, France, the Netherlands and the U.S.A. I have observed in the course of these that this new area of Diophantine approximation is arousing increasing interest. I hope that this volume may lend support to this and hopefully stimulate further research. Finally I would like to express my grateful thanks to the following: Professor Baker my research supervisor, St. John's College, the Science and Engineering Research Council, the Imperial Chemical Industries Educational Trust and Mrs. A.C. Williams the typist.

R.C.M.

Cambridge, 1983

CHAPTER I THE FUNDAMENTAL INEQUALITY

1 *INTRODUCTION*

This book demonstrates the applications of a fundamental inequality to the resolution of various general families of equations over function fields. The families concerned are the analogues of certain classical equations over number fields, and therefore we shall commence by discussing the researches of Thue, Mordell, Siegel and Baker on the theory of Diophantine equations. This discussion will be followed by an exposition of the problems surrounding the function field analogues, and an analysis of the contribution made towards this latter subject, principally by Osgood and Schmidt. The introduction will conclude with a summary of the results to be established herein, as consequences of the fundamental inequality.

The techniques of Diophantine approximation have long been applied to the study of general families of Diophantine equations. In his celebrated paper *[42]* of 1909, the Norwegian mathematician Thue employed approximation methods to prove that the equation $f(x,y) = 1$ has only finitely many solutions in integers x and y; here f denotes any irreducible binary form with rational coefficients and with degree at least three. The connexion between Thue equations, Diophantine approximation and transcendence will be discussed further in Chapter VI. The Johnian Mordell, in his Smith's prize essay of 1913 *[25]*, demonstrated that for certain values of the non-zero integer m, the equation $y^2 = x^3 + m$ has only finitely many integer solutions. Actually Mordell then conjectured that for certain other values of m there exist infinitely many integer solutions, but this was confounded by his own result *[27]* that the elliptic equation $y^2 = g(x)$ has only finitely many solutions in integers x, y, where g denotes a cubic polynomial with integer coefficients which contains no squared linear factor. The proof indirectly reduces the elliptic equation to a finite system of Thue equations, and thus the completion of the proof is provided by Thue's original theorem. His researches on the elliptic equation led Mordell to his famous "basis theorem" concerning the rational points on an elliptic curve *[28]*; this result was later generalised by Weil *[47]* to include curves of any positive genus, and the Mordell-Weil theorem has played a seminal role in this branch of number theory. In a letter to Mordell of 1925, later pseudonymously published *[39]*, Siegel proved that the hyperelliptic equation $y^2 = g(x)$ has only finitely many integer solutions; here g denotes a square-free polynomial of degree at least three and with integer coefficients. Although independent of that of Mordell, Siegel's method is to establish a direct connexion between the integer solutions of the hyperelliptic equation and the solutions of a finite set of Thue equations which are algebraic integers in a fixed number field. Siegel himself had already demonstrated *[38]* the generalisation of Thue's theorem required to complete the proof, that the equations arising possess but finitely many such solutions. In his illustrious paper of 1929 *[40]*, Siegel proved that any polynomial equation in two variables with integer coefficients has only finitely many solutions in integers, provided that the equation represents a curve of positive genus. The results discussed above of Thue, Mordell and Siegel himself are thus immediate corollaries to this general theorem. Furthermore, the theorem also applies when the curve associated with the equation has genus zero and possesses at least three infinite valuations and thereby provides a complete characterisation of those polynomial equations in two variables which have only finitely many integer solutions. The proof of Siegel's theorem employs two important results, the Mordell-Weil theorem and Siegel's previous result concerning the permissible degree of approximation of an algebraic number.

Unfortunately there is one serious reservation to be entertained towards all these results, namely that they are ineffective in failing to provide bounds on the solutions. Although they do establish that each of the various equations has only finitely many solutions, they furnish no means of actually determining the solutions. The problem of providing an effective analysis to families of Diophantine equations remained outstanding until 1968, when Baker, having established *[3]* a fundamental inequality relating to linear forms in the logarithms of algebraic numbers, applied this work to give a new proof *[4]* of Thue's theorem of 1909 which was effective. Baker's result thereby reduced to a finite amount of computation the problem of determining all the solutions in integers of the Thue equation $f(x,y) = 1$. Following this work, Baker employed the technique of Siegel *[39]* to establish *[6]* explicit bounds on integer solutions x, y of the hyperelliptic equation $y^2 = g(x)$; for the special case of the elliptic equation the method of Mordell served to provide him with sharper bounds *[5]*, and in fact it proved possible to resolve numerically certain equations, using these bounds together with some extensive machine computation. In 1970 Baker and Coates *[8]* discovered a simpler proof of the general theorem of Siegel which applies to any curve of genus one; the Riemann-Roch theorem serves to transform the equation to be solved into an elliptic equation. Their argument is effective and so provides an algorithm for the complete determination of all the integer solutions of any equation of genus one. Moreover, a similar method obtains when the associated curve has genus zero, and as usual possesses at least three infinite valuations. However, their proof does not appear to extend easily to equations of any higher genus, and an effective resolution of equations of genus two or more remains an important unsolved problem. In fact an even stronger result has very recently been discovered, for Faltings' resolution of the Mordell conjecture establishes that an equation of genus two or more has only finitely many rational solutions, not merely in integers as above. Unfortunately Faltings' method of proof is believed to be ineffective.

The subject of this book is the analogy for function fields of these results on Diophantine equations. We shall denote by k an algebraically closed field of characteristic zero, and by $k(z)$ the rational function field over k. Let us consider as an illustration the Thue equation $F(X,Y) = 1$, where F denotes a binary form with coefficients in $k(z)$ and possessing at least three distinct linear factors in some extension. Corresponding to the integers there is the polynomial ring $k[z]$, so the object of concern is the set of solutions X, Y in $k[z]$ of $F(X,Y) = 1$. In contrast to the classical Thue equation discussed above, there are four questions which may be posed concerning this equation and its polynomial solutions. Do these solutions have bounded degrees? If so, can their degrees be bounded effectively? Are there but finitely many solutions? Can all the solutions be determined effectively? This pluricity of problems results from the fact that the ground field k is infinite, and so bounds on the degrees of solutions do not of themselves enable the solutions to be determined, nor do they imply that only finitely many solutions exist. The first of these four questions was answered in the affirmative by Gill [13] in 1930, using an analogue of Thue's method relating to the approximation of algebraic functions by rational functions. As with Thue's work this result was ineffective, and in 1973 Osgood [31] established an effective version of Gill's theorem by a different method, relating to earlier researches of Kolchin [14] on algebraic differential equations. Osgood thereby resolved the second of the four questions, and his technique was further developed by Schmidt [35]; he calculated explicit bounds on the possible degrees of solutions. Schmidt also succeeded in solving the similar problem in the case of the hyperelliptic equation [36], and by extending this method further he was able to show [37] that bounds may be determined effectively in principle for all solutions in rational functions, not just in polynomials, of certain classes of hyperelliptic and superelliptic equations (see Chapter VIII), having previously done so explicitly for certain Thue equations. Concerning the third of the four questions, the criterion (Theorem 2) for the existence of infinitely many solutions may be deduced from the Manin-Grauert theorem [34] when F has at least four distinct linear factors. This celebrated theorem establishes for function fields the analogue of Faltings' theorem, that any equation of genus two or more may have only finitely many solutions in rational functions, provided that it is not birationally equivalent to an equation with coefficients in the ground field k. Algebraic geometry was highlighted in Grauert's work, and this approach continued with Shafarevich and Parsin; the latter gave bounds on the heights of these rational function solutions. Unfortunately these are multiply exponential and too large for any application.

In Chapter II we shall resolve the final, central and subsuming question of the four by establishing a simple algorithm for the determination of all the polynomial solutions of the Thue equation $F(X,Y) = 1$. The criterion for the existence of infinitely many solutions and a bound for the degrees of the solutions are both but immediate consequences of the analysis. Moreover, our approach is completely different from those of previous authors on the problem: the crux is an inequality, whose motivation lies in the study of linear forms in the logarithms of algebraic functions, and which therefore may be regarded as a parallel approach to Baker's to the Diophantine case above. This fundamental inequality also serves prominently in our solution of the hyperelliptic equation in Chapter III and of equations of genera zero and one in Chapter IV. Furthermore, by extending the application of the inequality to fields of positive characteristic, we shall succeed in Chapter VII in the complete resolution of the Thue and hyperelliptic equations in that circumstance; here the criteria for the existence of infinitely many solutions is necessarily more involved than in the case of characteristic zero, thereby illustrating the greater richness inherent over fields of positive characteristic. Finally, in Chapter VIII we shall further demonstrate the scope of the fundamental inequality by utilising it to yield explicit bounds on the solutions in rational functions, not merely in polynomials, of certain general classes of superelliptic equation. Throughout the strength of the fundamental inequality is reflected in the bounds calculated on the degrees of solutions, improving in each case on those of previous authors. This strength also appears in our explicit solutions of certain equations (II, §3; III, §4; VII, §2), carried out without resort to machine computation. This provides a stark contrast with the analysis for number fields, and provides further evidence of the efficacy of the inequality and the efficiency of the algorithms.

Our results in fact refer to a rather more general situation than that already described; actually we shall deal with the solutions integral over $k[z]$ in an arbitrary finite extension K, rather than just those in $k[z]$ itself. In §2 we shall recall some preliminary results on valuations and Puiseux expansions, requisite for the proof of the fundamental inequality (Lemma 2) in §3. We note that in order to speak meaningfully of the computation of solutions it is necessary to assume from the outset that the ground field k is presented explicitly [12]; this means here that there is an algorithm to determine the zeros of any polynomial with coefficients in k. We also remark that neither this assumption on k, nor even that of algebraic closure, is necessary for the validity of any of the bounds established here, provided that K does not extend the field of constants k. Finally we remark that, although the algorithms all apply to function fields in one variable, there is no difficulty in the extension to arbitrarily many variables, by means of a simple inductive argument.

2 PRELIMINARIES

In this section we shall introduce the concepts of a valuation on, a derivation on, and the genus of, a function field of one variable; these will serve as our basic tools throughout. Let us begin by recalling the definitions of the canonical valuations on the field $k(z)$. For each finite point a in k, every non-zero element f of $k(z)$ may be expanded as a formal Laurent series $\sum_{n=m}^{\infty} c_n(z-a)^n$, where each c_n is an element of k and c_m is non-zero. The additive valuation ord_a on $k(z)$ is then defined as $\mathrm{ord}_a f = m$. Similarly we may expand every non-zero f in $k(z)$ as a Laurent series in powers of $1/z$, and thence obtain the infinite valuation $\mathrm{ord}(f)$ on $k(z)$; $\mathrm{ord}(P) = -\deg(P)$ for P in $k[z]$. Then ord and, for each a in k, ord_a, define non-archimedean valuations on $k(z)$ with value group $\mathbb{Z}$; furthermore, *the sum formula*

$$\mathrm{ord}(f) + \sum_a \mathrm{ord}_a f = 0$$

is satisfied for each non-zero element f of $k(z)$. Finally we define $\mathrm{ord}(f) = \mathrm{ord}_a(f) = \infty$ if $f = 0$.

Let us recall also Puiseux's theorem (see *[11, Chapter III]*). We denote by

$$P(X,z) = X^d + P_1(z)X^{d-1} + \dots + P_d(z)$$

a monic polynomial in X with coefficients $P_1,\dots,P_d$ in $k(z)$. Then for each finite point a in k, the theorem asserts the existence of formal Puiseux series

$$y_{ij} = \sum_{h=m_i}^{\infty} c_{hi}\zeta_i^{hj}(z-a)^{h/e_{ai}} \qquad (1 \le j \le e_{ai},\ 1 \le i \le r_a)$$

such that

$$P(X,z) = \prod_{i=1}^{r_a} \prod_{j=1}^{e_{ai}} (X-y_{ij});$$

here the c_{hi} are elements of k and ζ_i is a primitive e_{ai}-th root of unity in k. Moreover, a similar assertion holds at ∞, with $1/z$ replacing $z-a$. An explicit construction of Puiseux expansions will be studied in further detail in Chapter V.

As in §1, K will denote some finite extension of $k(z)$, of degree d over $k(z)$, say. Since $K/k(z)$ is separable, it is generated by a single element, y say, whose minimal polynomial over $k(z)$ may be taken to be $P(X,z)$ as above. Then, for each a in k, and each i,j as above, the map $\sigma_{ij}: y \mapsto y_{ij}$ extends uniquely to give an embedding of K into the field of formal Puiseux expansions in fractional powers of z-a. Hence for each i, $1 \le i \le r_a$, we may construct a valuation v on K by defining

$$v(f) = \mathrm{ord}_v f$$

for f non-zero, where $\mathrm{ord}_v f$ denotes the order of vanishing of the Laurent series $\sigma_{ij}(f)$ in powers of *the local parameter* $z_v = (z-a)^{1/e_{ai}}$, where j is some suffix as above. We define *the ramification index* e_v of the valuation v to be $e_{ai} = v(z-a)$, and we write $v|a$ to denote that v extends the valuation $e_v \mathrm{ord}_a$ on $k(z)$; we also note that $\sum_{v|a} e_v = d$. Such valuations v, for any a in k, are termed *finite*, and they are characterised by the property $v(z) \ge 0$. A similar process may be applied to the infinite valuation ord, to construct valuations $v_1, \ldots, v_r$ on K. Each of these valuations v is defined as the order of vanishing of the Laurent expansion in powers of the local parameter $z_v = z^{-1/e_v}$; such valuations are called *infinite* and we write $v|\infty$, they are characterised by the property that $v(z) < 0$. We observe that all the additive valuations constructed above are non-archimedean with value group $\mathbb{Z}$ and so

$$v(fh) = v(f) + v(h) \ , \ v(f+h) \geq \min \{v(f), v(h)\}$$

for any f,h in K; furthermore the complete set of these valuations is actually independent of the initial choice of z in $K \setminus k$. Finally, we may characterise the non-zero elements of k by the property that $v(f) = 0$ for all v; the ring of elements of K integral over $k[z]$, which we term $\mathcal{O}$, is characterised by the property that $v(f) \ge 0$ for all finite valuations v.

The *norm* $N(f)$ of an element f in K is defined to be the product $\tau_1(f) \ldots \tau_d(f)$, where $\tau_1, \ldots, \tau_d$ denote distinct embeddings of K in some extension field of $k(z)$; since $N(f)$ lies in $k(z)$ it is actually independent of the choice of $\tau_1, \ldots, \tau_d$. Selecting for $\tau_1, \ldots, \tau_d$ the embeddings σ_{ij} defined above, we obtain

$$\mathrm{ord}_a N(f) = \sum_{v|a} v(f)$$

for each a in k, and f non-zero. Furthermore, the same result holds at ∞, and hence, applying the sum formula above to the non-zero element $N(f)$ of $k(z)$, we obtain *the sum formula* on K,

$$\sum_v v(f) = 0 \tag{1}$$

for any non-zero f in K, where the sum is taken over all the valuations v on K defined above: for each f only finitely many summands are non-zero.

We now wish to generalise the measure on $k[z]$ given by the degree of a polynomial. This is achieved by the *height* $H(f)$, defined by

$$H(f) = -\sum_v \min(0, v(f)) \tag{2}$$

for any f in K as in *[36]*; thus $H(f)$ is just the number of poles of f, counted according to multiplicity. We note that in view of the sum formula (1), $H(f) = \sum_v \max(0, v(f))$ for f non-zero, and that if in addition f lies in $k[z]$, then $H(f) = d.\deg f$. Thus in particular $H(z) = d$ and so, since the set of valuations is independent of the choice of z in $K \setminus k$, we have $H(f) = [K:k(f)]$ for any f in $K \setminus k$. We now wish to establish several elementary inequalities concerning the height function; these will be used throughout. First we have

$$\max\{H(f+h), H(fh)\} \le H(f) + H(h) \tag{3}$$

for any two elements f,h in K. We also wish to define the height of a polynomial with coefficients in K, and calculate a bound on the heights of the roots. Let us denote for any finite set S of elements of K

$$v(S) = \min_{s \in S} \{v(s)\} \quad \text{and} \quad H(S) = -\sum_v \min(0, v(S));$$

evidently

$$\max_{s \in S} \{H(s)\} \le H(S) \le \sum_{s \in S} H(s)$$

If P is a polynomial over K and S is the set of its coefficients, then the quantities $v(P)$ and $H(P)$ are defined to be $v(S)$ and $H(S)$ respectively. Suppose now that 1 lies in S, and μ is a non-zero element of K; we claim that

$$H(S) \le H(\mu S),$$

where μS is the set of elements μs for s in S. Thus replacing a polynomial by its monic equivalent suffers no increase in height. To establish the inequality, we observe that $v(\mu S) = v(\mu)+v(S)$ for each v, and so

$$H(\mu S) \ge -\sum_v v(\mu S) = -\sum_v v(\mu) - \sum_v v(S) = H(S)$$

as required, since by (1) $\sum_v v(\mu) = 0$, and $\sum_v v(S) = -H(S)$ as $1 \in S$. Now, Gauss' lemma yields

$$v(PQ) = v(P) + v(Q)$$

for any two polynomials P and Q; in particular, if P and Q are monic then

$$H(PQ) = H(P) + H(Q).$$

We thus conclude that if the non-zero polynomial $P(X)$ factorises in K as $f_o \prod_{i=1}^{n}(X-\alpha_i)$, then

$$H(f_o) \le H(P) \quad \text{and} \quad \sum_{i=1}^{n} H(\alpha_i) \le H(P). \tag{4}$$

We now introduce the global and local derivations on K, and consider the connexions between them; they will form the basis for the proof of the fundamental inequality in §3. The mapping $f \mapsto f'$ on $k(z)$, termed differentiation with respect to z, extends uniquely to a global derivation on K by virtue of the equation obtained from $P(y,z) = 0$,

$$y'\frac{\partial P}{\partial X}(y,z) + \frac{\partial P}{\partial z}(y,z) = 0.$$

Furthermore, for each valuation v on K we may define the local derivation $\frac{d}{dv}$ on the field of formal Laurent expansions in powers of z_v, namely differentiation with respect to z_v. Since $\frac{df}{dv} = f'\frac{dz}{dv}$ for each f in K, the sum formula (1) yields $\sum_v v(\frac{df}{dv}) = \sum_v v(\frac{dz}{dv})$ provided f' is non-zero. It is therefore permissible to define an invariant g of K/k by the equation

$$2g - 2 = \sum_v v(\frac{df}{dv}), \tag{5}$$

valid for any f in $K\setminus k$; g is termed the *genus* of K/k. Since the order of vanishing $v(\frac{df}{dv})$ is independent of the choice of local parameter at v, the genus is actually independent of the choice of z in $K\setminus k$. (It follows from the Riemann-Roch theorem that g is non-negative, and for convenience only we shall assume this throughout. For example, without assuming $g\geq 0$ we obtain the bound $H(X,Y)\leq\max(5H,8H+2g+r-2)$ in place of Theorem 3.) We shall record here two inequalities on the order of vanishing of the local derivative of f at v; these are readily obtained by considering the Laurent expansion of f in powers of z_v. We have namely

$$v(\frac{df}{dv}) \geq v(f)-1 \text{ for any f in K, and } v(\frac{df}{dv})\geq 0 \text{ if } v(f)\geq 0. \tag{6}$$

These inequalities, together with the genus formula (5), form the crux of the proof of the fundamental inequality in the next section.

Later it will be necessary to extend the field K, and so here we investigate briefly the behaviour of valuations, the height and the genus under such an extension. Let us denote by L a finite extension of K, of degree δ over K, say. Valuations w on L may be constructed as above; each such w, upon restriction to K, forms a valuation with value group $\varepsilon_w\mathbb{Z}$ for some $\varepsilon_w>0$. In this case we write $w|v$, where $w=\varepsilon_w v$, and ε_w is termed the *relative ramification index* of w over v. In view of Puiseux's theorem, upon comparing degrees we have $\sum_{w|v}\varepsilon_w=\delta$ for each v. Furthermore, if f is an element of $K\setminus k$ then

$$H_L(f) = [L:k(f)] = [L:K][K:k(f)] = \delta H(f),$$

where $H_L(f)$ denotes the height of f in L; if f lies in k then $H_L(f)=\delta H(f)$ also, since both sides are zero. Finally we wish to relate the genus g_L of L to that of K. Since $z_v=z_w^{\varepsilon_w}$ for $w|v$, we obtain $w(\frac{dz_v}{dz_w}) = \varepsilon_w-1$, and so $w(\frac{df}{dw}) = w(\frac{df}{dv})+(\varepsilon_w-1) = \varepsilon_w v(\frac{df}{dv})+(\varepsilon_w-1)$ for any f in K. Hence by (5) we deduce that

$$(2g_L-2) - \delta(2g-2) = \sum_w (\varepsilon_w-1). \tag{7}$$

We conclude the discussion of preliminaries and this section by establishing a lemma which asserts that an element of K may be determined effectively from the set of values it assumes. This lemma will form an essential tool in the actual computation of solutions. Further, in Chapter V it is employed for the construction of particular rational functions on curves of small genus; there explicit estimates are made for each stage of this proof.

Lemma 1

Suppose that for each valuation v *on* K *we are given an integer* m_v*, such that* m_v *is non-zero for at most finitely many* v*. Then we can determine effectively whether there exists an element* f *in* K *such that* $v(f)=m_v$ *for all* v*. Moreover, if such an* f *exists then it may be computed, and it is unique to within a non-zero factor in* k.

Proof. We recall that y is a primitive element of the extension K/k(z); we may assume that in fact y lies in $\mathcal{O}$. Any element f of K may now be expressed uniquely in the form

$$f = \sum_{i=1}^{d} h_i(z)y^{i-1},$$

with $h_1,\ldots,h_d$ in k(z). If $\tau_1,\ldots,\tau_d$ denote distinct embeddings of K in some extension field of k(z), then the equations

$$\tau_j(f) = \sum_{i=1}^{d} h_i(z) \ (\tau_j(y))^{i-1} \qquad (1\le j\le d)$$

serve to express each of $h_1D,\ldots,h_dD$ as polynomials in the $\tau_j(f)$ and $\tau_j(y)$. Here D denotes the determinant of order d with entries $(\tau_j(y))^{i-1}$, so $D = \prod_{i\neq j}(\tau_i(y)-\tau_j(y))$, and is thus an element of k[z], independent of the particular choice of the embeddings $\tau_1,\ldots,\tau_d$. By selecting $\tau_1,\ldots,\tau_d$ to be the σ_{ij} defined above, we deduce that if $v(f)=m_v$ for each v, then for each a in k there is an integer t_a such that $\mathrm{ord}_a h_i \ge t_a$ for $1\le i\le d$; furthermore, t_a may be assumed to be non-zero for at most finitely many a. Hence if Q denotes the rational function $\prod_{a\in k}(z-a)^{t_a}$, then $\text{ord}_a(h_i/Q) \geq 0$ for each a in k, $1 \leq i \leq d$, and so $P_i = h_i/Q$ is a polynomial in $k[z]$. By selecting $\tau_1, \ldots, \tau_d$ to be the embeddings corresponding to the infinite valuations on K, we deduce that there is an integer t such that $\text{ord} h_i \geq t$ for $1 \leq i \leq d$, and so each P_i has degree at most $J = \text{ord}Q - t$. Thus

$$f/Q = \sum_{j=0}^{J} \sum_{i=1}^{d} a_{ij} z^j y^{i-1} \tag{8}$$

for some elements a_{ij} in k to be determined. We now show that we may replace the infinite system of equations $v(f) = m_v$ by a finite system of inequalities, namely $v(f) \geq m_v$ for each v such that $v|\infty$ or $m_v > v(Q)$. For we may assume, by the sum formula (1), that $\sum_v m_v = 0$, and so any f given by (8) which also satisfies these inequalities has f=0 or

$$0 = \sum_v v(f/Q) \geq \sum_v (m_v - v(Q)) = 0,$$

since $v(f/Q) \geq 0$ for all $v \nmid \infty$. Hence if f is non-zero then equality holds throughout and $v(f) = m_v$ for each valuation v, as required. Now the finite system of inequalities $v(f) \geq m_v$ presented above is equivalent, by consideration of the Puiseux expansion of y at each such v, to a finite system of linear equations in the a_{ij}, with coefficients in k. We may determine effectively whether a system of linear equations over k has a non-zero solution, and we may compute a solution if it exists. Hence we may determine whether there exists an element f in K such that $v(f) = m_v$, and, if so, such an element may be computed. Finally, if f_1 and f_2 are two elements of K with $v(f_1) = m_v = v(f_2)$ for all v, then $v(f_1/f_2) = 0$ and so f_1/f_2 is a non-zero element of k. Hence f is unique to within a non-zero factor in k, and thus the proof of the lemma is complete.

3 *THE FUNDAMENTAL INEQUALITY*

In this final section of Chapter I we shall establish an inequality (Lemma 2) which will serve as our major weapon in attacking the various problems presented in this book. The inequality may be regarded as a form of lower bound for linear forms in the logarithms of algebraic functions, but we avoid the actual use of such logarithms, and the consequent discussion of units in function fields. The bounds derived from the inequality for the heights of solutions of the various equations in each case vary only as a linear function of the height of the equation concerned, and are not exponential as in Baker's work *[4]* on the classical case of algebraic numbers. Our analysis is in fact achieved by a consideration of the connexion between the local and global derivations on K, as expressed in the genus formula (5). In particular it may be observed that, unlike Schmidt *[36]*, we employ neither algebraic differential equations nor even the Riemann-Roch theorem in our analysis. In Chapter II we shall apply our inequality to solve the Thue equation, and in the succeeding chapters it will be seen that the inequality also plays crucial roles in the solutions of each of the hyperelliptic, genus zero and genus one families of equations. Furthermore, it is the generalisation of the inequality to fields of positive characteristic which forms the starting point for our complete resolution of the Thue and hyperelliptic equations in that circumstance. Finally, the inequality also serves in Chapter VIII to provide a bound on all the solutions in K, not just in $\mathcal{O}$, of some superelliptic and Thue equations.

In order to show how the necessity for such an inequality arises, let us commence with a brief discussion of the Thue equation. Suppose, then, that $\alpha_1,\ldots,\alpha_n,\mu$ are elements of $\mathcal{O}$ with μ non-zero and $\alpha_1,\alpha_2,\alpha_3$ distinct. We shall be concerned in Chapter II with the effective determination of all the solutions X,Y in $\mathcal{O}$ of the Thue equation

$$(X - \alpha_1 Y) \ldots (X - \alpha_n Y) = \mu \tag{9}$$

We shall write, for brevity, $\beta_i = X-\alpha_i Y$ for $1 \le i \le n$, so that each β_i is an element of $\mathcal{O}$. Furthermore, $\beta_1 \ldots \beta_n = \mu$, so each μ/β_i is also an element of $\mathcal{O}$, and so, if v is any finite valuation on K such that $v(\mu)=0$, then $v(\beta_i)=0$, $1 \le i \le n$. Also, for any three suffixes i,j,ℓ, we have

$$\beta_i(\alpha_j-\alpha_\ell) + \beta_j(\alpha_\ell-\alpha_i) + \beta_\ell(\alpha_i-\alpha_j) = 0; \tag{10}$$

it is this identity which forms the crux of the analysis, both here and in the classical case of algebraic numbers by Baker *[4]*. In the classical case each β_i is factorised as a product of a unit and an element of bounded height; this is possible since each β_i divides the fixed element μ. This factorisation transforms (10) into the equation

$$\psi\eta_1^{b_1} \cdots \eta_s^{b_s} = 1 + \omega(\beta_j/\beta_i),$$

where $\eta_1,\ldots,\eta_s$ form a basis for the unit group of the given algebraic number field, $b_1,\ldots,b_s$ are integers, and ψ,ω have bounded heights. By taking the logarithm of the left hand side and applying a lower bound for the resulting linear form, an upper bound on the absolute value of β_i/β_j is obtained. This bound suffices for the determination of all integral solutions of the Thue equation in the given algebraic number field.

The lemma proved here refers to an equation of the same type as (10), and asserts an upper bound on the height of β_i/β_j, under the sole assumption that the values of $\beta_i,\beta_j,\beta_\ell$ are restricted at all but finitely many valuations. The corollary proved below shows that, in addition, the full range of possibilities for β_i/β_j may be determined effectively. This type of equation, with restrictions on the values of the summands, will be seen to arise naturally for the hyperelliptic, genus zero and superelliptic equations also.

Lemma 2

Suppose that γ_1, γ_2 *and* γ_3 *are non-zero elements of* K *with* $\gamma_1+\gamma_2+\gamma_3=0$, *and such that* $v(\gamma_1) = v(\gamma_2) = v(\gamma_3)$ *for each valuation* v *not in the finite set* V. *Then either* γ_1/γ_2 *lies in* k, *in which case* $H(\gamma_1/\gamma_2) = 0$, *or*

$$H(\gamma_1/\gamma_2) \le |V| + 2g - 2, \tag{11}$$

where $|V|$ *denotes the number of elements of* V.

Proof. It will be convenient to write $f = \gamma_1/\gamma_2$, so in view of the equation $\gamma_1+\gamma_2+\gamma_3=0$ we have $\gamma_3/\gamma_2 = -(f+1)$; it may be assumed that f does not lie in k. Let V_1, V_2 and V_3 denote the set of valuations v at which $v(f) < 0$, $v(f) > 0$ or $v(f+1) > 0$ respectively, so that each of the disjoint sets V_1, V_2, V_3 is contained in V. Furthermore, for v in V_1, V_2 or V_3, let n_v denote the positive integer $-v(f)$, $v(f)$ or $v(f+1)$ respectively. In view of the sum formula (1), applied to f and to f+1, we have $H(f) = \sum n_v$, whether the sum is taken over v in V_1, V_2 or V_3. We now employ the inequalities (6) enunciated in §2 to estimate the order of vanishing of each of the local derivatives of f. If v lies in V_1, then

$v(\frac{df}{dv}) \geq v(f)-1 = -n_v-1$, whilst if v lies in V_2 or V_3, then $v(\frac{df}{dv}) \geq n_v-1$. Since $v(\frac{df}{dv}) \geq 0$ for all other valuations v, the genus formula (5) yields

$$2g - 2 \geq - \sum(n_v+1) + \sum(n_v-1),$$

where the first sum is taken over valuations v in V_1, and the second over v in $V_2 \cup V_3$. Since each of V_1, V_2 and V_3 is contained in V, the required inequality (11) now follows.

Corollary

Under the same assumptions made in Lemma 2, then either γ_1/γ_2 *lies in* k, *or it has only finitely many possibilities in* K, *which may be determined effectively.*

Proof. It follows from the hypotheses of Lemma 2 that $v(\gamma_1/\gamma_2) = 0$ for each valuation v not in the finite set V. Furthermore, it follows from the conclusion, (11), that $H(\gamma_1/\gamma_2) = -\sum_v \min(0, v(\gamma_1/\gamma_2))$ is bounded. Since by the sum formula (1) $H(\gamma_1/\gamma_2)$ is also equal to $\sum_v \max(0, v(\gamma_1/\gamma_2))$, we conclude that the complete set of values $v(\gamma_1/\gamma_2)$, as v runs over the valuations on K, has only a finite number of possibilities. We may now apply Lemma 1, and deduce that, up to a non-zero factor in k, γ_1/γ_2 has only a finite number of possibilities, which are effectively determinable. Thus we may write $\gamma_1/\gamma_2 = a\alpha$, where a is some non-zero element of k and α belongs to a finite computable subset of K. However, the same analysis may be applied to the ratio γ_3/γ_2, and which therefore may be written in the form $b\beta$, where b is some non-zero element of k and β belongs to a finite computable subset of K. For each of the finitely many pairs α, β we wish to determine the possible values for a,b in k. But $\gamma_1+\gamma_2+\gamma_3 = 0$ yields $a\alpha+b\beta = -1$, which, upon differentiating, gives $a\alpha'+b\beta' = 0$. We conclude that, unless α' and β' both vanish, that is, unless α and β actually lie in k, then a and b are determined. Thus we have proved that unless γ_1/γ_2 lies in k, it has only a finite number of effectively determinable possibilities in K, as required.

In the succeeding chapters Lemma 2 and its corollary will be applied to the various families of equations under consideration, commencing in Chapter II with the Thue equation. Lemma 2 itself will be employed for the estimation of bounds on the heights of the solutions, whilst its corollary will be the principal tool in the construction of algorithms for the actual determination of solutions.

CHAPTER II THE THUE EQUATION

1 THE SOLUTION

In this chapter we shall make the first application of our fundamental inequality, which is to the Thue equation *[21]*. This first section will be devoted to the construction of an algorithm for the effective determination of all the solutions of a general Thue equation over an arbitrary finite extension K of k(z). The following section 2 will be concerned with the establishment of a bound on the heights of the solutions; this bound represents an improvement on all those previously obtained *[21,35,36]*. Finally, in §3 we shall illustrate our algorithm for the solution by solving completely a particular Thue equation over a function field of genus one. The example provides an illustration of the variety of solutions that may arise in this context. Furthermore, it will be seen that no machine computation is used; this is a further reflection on the efficiency of the algorithm and the sharpness of the bounds, both of which derive from the fundamental inequality.

Let us now assume, as in §3 of Chapter I, that $\alpha_1,\ldots,\alpha_n$ and μ are elements of $\mathcal{O}$ such that α_1, α_2 and α_3 are distinct and μ is non-zero. We shall prove the following theorems in this section.

Theorem 1

All the solutions X,Y *in* $\mathcal{O}$ *of the Thue equation*

$$(X - \alpha_1 Y) \ldots (X - \alpha_n Y) = \mu \tag{9}$$

may be determined effectively.

Theorem 2

The equation (9) *has an infinite number of solutions in* $\mathcal{O}$ *if and only if there exists a substitution*

$$X = \alpha x + \beta y , \; Y = \gamma x + \delta y \qquad (\alpha,\beta,\gamma,\delta \in \mathcal{O}, \alpha\delta \neq \beta\gamma)$$

which transforms the left hand side of (9) *into* $\mu F(x,y)$ *where* F *is a binary form with coefficients in the ground field* k. *Moreover, if such a substitution exists then there will be the infinite family derived from the solutions* x, y *in* k *of* $F(x,y) = 1$, *but otherwise only finitely many solutions* X, Y *in* $\mathcal{O}$ *of* (9).

Theorem 3

All the solutions X, Y *in* $\mathcal{O}$ *of* (9) *satisfy*

$$H(X,Y) \leq 8H + 2g + r-1;$$

here H *denotes the height of the polynomial* $(x-\alpha_1)\ldots(x-\alpha_n)/\mu$, g *denotes the genus of* K/k, *and* r *denotes the number of infinite valuations on* K.

The bound in Theorem 3 may be compared with one previously obtained *[36]* using different methods, namely $89H + 212g + r-1$. Furthermore, the bound is linear with respect to the height of the equation, in contrast with the classical case when the bounds obtained by Baker are of exponential complexity *[4]*: the comparison is quite legitimate, even though the direct analogue of the classical height function is $\exp(H(f)/[K:k(f,z)])$ rather than $H(f)$.

The remainder of this section is devoted to the exposition of the proofs of Theorems 1 and 2. Let us write, for brevity, $\beta_i = X - \alpha_i Y$ for $1 \leq i \leq n$, where X, Y denotes some solution in $\mathcal{O}$ of (9). We then obtain $\beta_1 \ldots \beta_n = \mu$ and, for any three suffixes i, j, ℓ,

$$\beta_i(\alpha_j-\alpha_\ell) + \beta_j(\alpha_\ell-\alpha_i) + \beta_\ell(\alpha_i-\alpha_j) = 0. \qquad (10)$$

As disclosed in the last section, it is this identity to which our fundamental inequality may be referred; the restriction on the values of the summands required in its hypotheses being derived from the relation $\beta_1 \ldots \beta_n = \mu$. Let us first, however, observe that it in fact suffices, for the determination of all solutions X, Y in $\mathcal{O}$ of (9), to establish the full range of possibilities for any one of the ratios β_i/β_j with $\alpha_i \neq \alpha_j$. For the ratio X/Y may be expressed as $(\alpha_i\beta_j-\alpha_j\beta_i)/(\beta_i-\beta_j)$ and so, by virtue of the equation (9), X and Y are determined, up to a factor of an n-th root of unity in k, by such a ratio β_i/β_j. Now, if v is a finite valuation on

K, then since each β_i lies in $\mathcal{O}$ we have $v(\beta_i) \geq 0$, and so since $\beta_1 \ldots \beta_n = \mu$ either $v(\mu) > 0$ or $v(\beta_i) = 0$ for $i=1,\ldots,n$. Thus if α_i, α_j, α_ℓ are distinct, let us denote by V the set of valuations v such that $v|\infty$, $v(\mu) > 0$, or $v(\alpha_j-\alpha_\ell)$, $v(\alpha_\ell-\alpha_i)$, $v(\alpha_i-\alpha_j)$ are not all equal. Hence V is a finite computable set, and if we write $\gamma_1 = \beta_i(\alpha_j-\alpha_\ell)$, $\gamma_2 = \beta_j(\alpha_\ell-\alpha_i)$, $\gamma_3 = \beta_\ell(\alpha_i-\alpha_j)$, then from (10) we have $\gamma_1 + \gamma_2 + \gamma_3 = 0$, and, by the definition of V we have $v(\gamma_1) = v(\gamma_2) = v(\gamma_3)$ for all v not in V. We may therefore conclude from the corollary to Lemma 2 that γ_1/γ_2 is either an element of k, or belongs to a finite effectively determinable subset of K.

We shall now distinguish between, and treat separately, two different types of Thue equation, depending on the parameters $\alpha_1,\ldots,\alpha_n$. Let us denote by λ_i the cross-ratio $(\alpha_3-\alpha_i)(\alpha_1-\alpha_2)/(\alpha_i-\alpha_2)(\alpha_3-\alpha_1)$ for $1 \leq i \leq n$, where we write $\lambda_i = \infty$ if $\alpha_i = \alpha_2$. In the first case we make the assumption that some finite cross ratio λ_i is not an element of k; we proceed to show that in this case the Thue equation (9) has necessarily only finitely many solutions X, Y in $\mathcal{O}$. For the quotient of the two ratios $\beta_2(\alpha_3-\alpha_i)/\beta_3(\alpha_i-\alpha_2)$ and $\beta_2(\alpha_3-\alpha_1)/\beta_3(\alpha_1-\alpha_2)$ is λ_i, and so it follows from the assumption that at least one of these two ratios is not an element of k. In view of the previous paragraph we deduce that β_2/β_3 has only a finite number of possibilities, effectively determinable. But β_2/β_3 determines X and Y, up to a factor of an n-th root of unity in k, and so all the solutions in $\mathcal{O}$ of the Thue equation (9) may be computed, and these solutions are finite in number. The proofs of Theorems 1 and 2 are thus complete in this case, when some finite cross-ratio λ_i is not in k.

Let us henceforth assume that the alternative case holds, so that all the finite cross-ratios λ_i are elements of k. Here we shall transform the Thue equation by means of the substitution

$$X = \alpha_3\rho x + \alpha_2\tau y \; , \; Y = \rho x + \tau y, \tag{12}$$

where

$$\rho = \pi \frac{\alpha_1-\alpha_2}{\alpha_3-\alpha_2} \; , \; \tau = \pi \frac{\alpha_3-\alpha_1}{\alpha_3-\alpha_2} \; ,$$

and π is some non-zero element of K yet to be specified. If α_i differs from both α_1 and α_2, then $X - \alpha_i Y = \pi(\alpha_1-\alpha_i)(u_i x + v_i y)$, where

$u_i = \lambda_i/(\lambda_i-1)$ and $v_i = 1/(1-\lambda_i)$. Furthermore, $X - \alpha_2 Y = \pi(\alpha_1-\alpha_2)x$, so we choose $u_i = 1$, $v_i = 0$ if $\alpha_i = \alpha_2$; $X - \alpha_1 Y = \pi(\alpha_1-\alpha_2)(\alpha_3-\alpha_1)(\alpha_3-\alpha_2)^{-1}(x-y)$, so we choose $u_i = -v_i = 1$ if $\alpha_i = \alpha_1$. The substitution (12) transforms the Thue equation (9) into

$$\pi^n \theta \prod_{i=1}^{n} (u_i x + v_i y) = \mu , \tag{13}$$

where θ is some element of K depending only on $\alpha_1,\ldots,\alpha_n$, and u_i, v_i lie in k for $1 \le i \le n$. Furthermore, $\beta_2(\alpha_3-\alpha_1)/\beta_3(\alpha_1-\alpha_2)$ is equal to $-x/y$, and so we deduce from the corollary to Lemma 2 that either x/y lies in k, or it belongs to some finite computable subset of K. Since x/y determines β_2/β_3 uniquely, we conclude as above that x/y determines X and Y, up to a factor of an n-th root of unity. Thus each of the finitely many possibilities for x/y in $K \setminus k$ may be examined and dealt with in turn. We now assume that x/y is an element of the ground field k, and we wish to determine its range of possible values. First we observe that, since u_i and v_i lie in k for $1 \le i \le n$, so also does $y^{-n} \prod_{i=1}^{n} (u_i x + v_i y)$. Thus if (13) has any solutions x, y with x/y in k, then μ/θ must be an n-th power of some element in K. Lemma 1 enables us to determine effectively whether this is indeed the case; if so, then we will choose π such that $\pi^n = \mu/\theta$. The left hand side of (13) is now equal to $\mu F(x,y)$, where $F(x,y) = \prod_{i=1}^{n} (u_i x + v_i y)$ is a binary form with coefficients in the ground field k. Since x/y is an element of k and $F(x,y) = 1$, it follows that x and y are both elements of k. Furthermore, if x and y are elements of k satisfying $F(x,y) = 1$, then X and Y, given by the transformation (12), will be elements of K satisfying the Thue equation (9). It remains only to be determined when X and Y actually lie in $\mathcal{O}$ if x and y lie in k; the proofs of Theorems 1 and 2 will then be complete.

Let us suppose first that ρ and τ both lie in $\mathcal{O}$. In this case all four coefficients of the substitution (12) are elements of $\mathcal{O}$, and x, y may assume any values in k such that $F(x,y) = 1$; the corresponding solution X, Y of (9) will then lie in $\mathcal{O}$. Furthermore, we have already observed that there are only a finite number of possibilities for x/y in $K \setminus k$. Thus we conclude that, apart from the infinite family derived from the solutions in k of $F(x,y) = 1$, the Thue equation (9) has only finitely many solutions X, Y in $\mathcal{O}$. The proofs of Theorems 1 and 2 are thus complete in this case also.

Finally let us suppose that ρ and τ do not both lie in $\mathcal{O}$. We shall complete the proofs of Theorems 1 and 2 in this final case by showing that x/y has at most one possibility in k. This, together with the finitely many possibilities in $K \setminus k$, leads to the complete set of solutions in $\mathcal{O}$ of the Thue equation (9), which are then finite in number. For some finite valuation v on K, we have $v(\rho) < 0$ or $v(\tau) < 0$ by assumption. If we expand ρ and τ as a Laurent series in powers of the local parameter z_v, we obtain

$$\sigma_v(\rho) = \sum_{h=v(\rho)}^{\infty} a_h z_v^h , \quad \sigma_v(\tau) = \sum_{h=v(\tau)}^{\infty} b_h z_v^h ,$$

where a_h and b_h lie in k, and $a_{v(\rho)} \neq 0 \neq b_{v(\tau)}$. However, if x/y lies in k and $F(x,y) = 1$ then x and y both lie in k, and so the expansion of Y at v is equal to $\sum_h (a_h x + b_h y) z_v^h$. Since Y lies in $\mathcal{O}$ we have $v(Y) \geq 0$ and so $a_h x + b_h y = 0$ for $h<0$. Thus the ratio x/y is determined, as required, since this equation is non-trivial for $h = v(\rho,\tau) < 0$.

The proofs of Theorems 1 and 2 are now complete, and we deduce that there are four possibilities for the set of solutions in $\mathcal{O}$ of the Thue equation (9), namely no solutions, a finite number of solutions, a single infinite family of solutions, and a single infinite family together with an isolated finite set of solutions. Examples of equations of the first three types are readily found; in particular we have the following.

(i) $X(X-Y)(X+Y) = z$ has no solution in $k[z]$. For if there were a solution, two of the polynomials X, X-Y, X+Y would have degree 0 and the third would have degree 1, which is impossible.

(ii) $X(X-Y)(X-zY) = 1$ has a finite number of solutions in $k[z]$. For each of X, X-Y, X-zY must lie in k, so Y=0 and $X^3=1$.

(iii) $X(X-Y)(X+Y) = 1$ has a single infinite family of solutions in $k[z]$. For X and Y must lie in k, and X may assume any non-zero value.

An example of the fourth type is presented in the concluding section of this chapter.

2 *BOUNDS*

This section will be devoted to a proof of Theorem 3, that all the solutions X, Y in $\mathcal{O}$ of the Thue equation (9) satisfy

$$H(X,Y) \leq 8H + 2g + r - 1 ,$$

where H is the height of the polynomial $(x-\alpha_1)\ldots(x-\alpha_n)/\mu$, g is the genus of K/k and r is the number of infinite valuations on K. Schmidt has obtained the bound *[36]* 89H + 212g + r - 1 for H(X,Y), under slightly different conditions on $\alpha_1,\ldots,\alpha_n$.

We shall first employ our fundamental inequality (Lemma 2) to deduce an upper bound on the height of a certain element of K. The proof of Theorem 3 will then consist of two parts: first, an estimation of the size of the set V in (11), and secondly, a deduction of the required bound on H(X,Y). Let us assume that α_i, α_j, α_ℓ are distinct, so we may rewrite the identity (10) as

$$\beta_i h + \beta_j - \beta_\ell(h+1) = 0 ,$$

where $h = (\alpha_j-\alpha_\ell)/(\alpha_\ell-\alpha_i)$. Let us also denote by V the set of valuations v on K such that $v|\infty$, $v(\mu)>0$, $v(h)>0$, $v(h+1)>0$, or $v(h)<0$, as in §1. If v is a finite valuation on K, then $v(\beta_i)\geq 0$, $1\leq i\leq n$, since each β_i lies in $\mathcal{O}$. Hence if in addition $v(\mu) = 0$, then $v(\beta_i) = 0$, $1\leq i\leq n$, and so if v lies outside V, then $v(\beta_i h) = v(\beta_j) = v(\beta_\ell(h+1)) = 0$ as required. The fundamental inequality (11) now yields $H(f) = 0$ or $H(f) \leq |V|+2g-2$, and so

$$H(f) \leq |V| + 2g - 1 ,$$

where $f = \beta_i h/\beta_j$; this follows since $|V| \geq 1$ and $g \geq 0$. Now, if v is a finite valuation such that $v(h)>0$, then $v(\alpha_j-\alpha_\ell)>0$ since $v(\alpha_\ell-\alpha_i)\geq 0$. Similarly, if $v\nmid\infty$ and $v(h)<0$ or $v(h+1)>0$, then $v(\alpha_\ell-\alpha_i)>0$ or $v(\alpha_j-\alpha_i)>0$. We deduce that

$$|V| \leq \sum_{v|\infty} 1 + \sum_{v} \max(0,\ v(\mu),\ v(\alpha_j-\alpha_\ell),\ v(\alpha_\ell-\alpha_i),\ v(\alpha_j-\alpha_i))$$

$$\leq r + H(\mu) + H(\alpha_j-\alpha_\ell) + H(\alpha_\ell-\alpha_i) + H(\alpha_j-\alpha_i).$$

The initial discussion on heights in Chapter I provides the inequalities

$H(\mu) \leq H$ and $\sum_{i=1}^{n} H(\alpha_i) \leq H$, and (3) gives $H(\alpha_j-\alpha_\ell) \leq H(\alpha_j) + H(\alpha_\ell)$, whence $|V| \leq r + 3H$, and so

$$H(f) \leq 3H + 2g + r - 1.$$

We now wish to establish the required bound on $H(X,Y)$. Let us first observe that $\beta_i/\beta_j = f/h$ and $\beta_\ell/\beta_j = (f+1)/(h+1)$, whence

$$-\min(0,v(\beta_i/\beta_j),v(\beta_\ell/\beta_j)) \leq -\min(0,v(f)) + \max(0,v(h)) + \max(0,v(h+1)).$$

Summing this last relation over the valuations v on K, we obtain

$$H(\beta_i/\beta_j,\beta_\ell/\beta_j) \leq H(f) + 2H(h),$$

since $H(h) = \sum_v \max(0, v(h)) = \sum_v \max(0, v(h+1))$. Now $h = (\alpha_j-\alpha_\ell)/(\alpha_\ell-\alpha_i)$, so (3) gives

$$H(h) \leq H(\alpha_j-\alpha_\ell) + H(\alpha_\ell-\alpha_i) \leq H + H(\alpha_\ell).$$

Also $h+1 = (\alpha_j-\alpha_i)/(\alpha_\ell-\alpha_i)$, so $H(h) = H(h+1) \leq H + H(\alpha_i)$, and thus $2H(h) \leq 3H - H(\alpha_j)$. Combining this with the above we obtain

$$H(\beta_i/\beta_j,\beta_\ell/\beta_j) \leq 6H - H(\alpha_j) + 2g + r - 1,$$

provided α_i, α_j and α_ℓ are distinct. However, this inequality is evidently also valid if some of α_i, α_j and α_ℓ coincide. Hence we obtain

$$H(\beta_i^n/\mu,\beta_\ell^n/\mu) \leq \sum_{j=1}^{n} H(\beta_i/\beta_j,\beta_\ell/\beta_j) \leq (6n-1)H + n(2g+r-1).$$

Since $H(\beta_i^n,\beta_\ell^n) = n\,H(\beta_i,\beta_\ell)$, this yields

$$H(\beta_i,\beta_\ell) \leq 6H + 2g + r - 1.$$

However, we may express X and Y in terms of β_i and β_ℓ, namely

$$X = (\alpha_\ell\beta_i-\alpha_i\beta_\ell)/(\alpha_\ell-\alpha_i), \quad Y = (\beta_i-\beta_\ell)/(\alpha_\ell-\alpha_i),$$

from which we deduce that

$$H(X,Y) \leq H(\beta_i,\beta_\ell) + H(\alpha_\ell) + H(\alpha_i) + H(\alpha_\ell-\alpha_i).$$

Since $H(\alpha_\ell-\alpha_i) \leq H(\alpha_\ell) + H(\alpha_i) \leq H$, the required result now follows.

3 AN EXAMPLE

In this section we shall determine all the integral solutions of the Thue equation

$$X(X - Y)(X + Y) = 1$$

in the field K, generated over k(z) by y and t satisfying

$$y^2 = t^3 - 1 \quad , \quad t^2 + tz + 1 = 0.$$

This example illustrates the general method of solution described in §1, and will also furnish an instance of an equation which possesses an infinite family of solutions, namely those with X, Y in k, together with finitely many other solutions in *O*. This typical example confirms the efficiency of the general algorithm, since no machine computation is used.

The Puiseux expansions of y about ∞ may be calculated readily, and are

$$\sigma_1(y) = \chi(z_1^{-3} - \tfrac{3}{2}z_1 + \ldots) \quad , \sigma_2(y) = \chi(1 + \tfrac{1}{2}z_2^3 + \ldots) \; ,$$

$$\sigma_3(y) = -\chi(1 + \tfrac{1}{2}z_3^3 + \ldots),$$

where χ is an element of k with $\chi^2 = -1$, and the local parameter z_i is equal to $z^{-\frac{1}{2}}$ when i = 1, and z^{-1} when i = 2, 3. Hence K has degree 4 over k(z), and there are three infinite valuations on K with ramification indices 2, 1, 1 respectively. Later we shall need the Puiseux expansions of t, which we record here as

$$\sigma_1(t) = -z_1^{-2} + z_1^2 + z_1^6 + \ldots \quad , \sigma_2(t) = -z_2 - z_2^3 - \ldots \quad ,$$

$$\sigma_3(t) = -z_3 - z_3^3 - \ldots \quad .$$

We now wish to calculate the genus of K/k, and to do this we apply the genus formula (7),

$$2g - 2 = \sum_{v}(e_v - 1) - 2d,$$

since k(z)/k has genus 0. By considering the Puiseux expansions of y at valuations $v|a$ with $a = \pm 2, \pm 1$, it may be established that the sum $\sum_v (e_v - 1)$ over such valuations v is equal to 7. We now maintain that for other finite valuations v, $e_v = 1$, and so K has genus 1. For by direct calculation we have $D = 16(z-2)^2(z+2)^3(z-1)^6(z+1)^4$, where $D = \prod_{i \neq j}(\tau_i(y) - \tau_j(y))$, as defined in Lemma 1. It therefore suffices to prove that if v is finite and satisfies $e_v > 1$, then $v(D) > 0$. For if σ_{ij} is an embedding corresponding to the valuation v, so that $\sigma_{ij}(y) = \sum_{h=0}^{\infty} c_h z_v^h$, then $\sigma_{ij'}(y) = \sum_{h=0}^{\infty} c_h \zeta^h z_v^h$, where j' is a suffix different from j with $1 \leq j' \leq e_v$, and $\zeta \neq 1$ is an e_v-th root of unity. Thus $v(\sigma_{ij}(y) - \sigma_{ij'}(y)) > 0$, and so, since we may select for $\tau_1, \ldots, \tau_d$ the embeddings corresponding to the valuations $w|a$ for some fixed a in k, and since y lies in $\mathcal{O}$, we conclude that $v(D) > 0$ as required. Less mechanically, although less generically, we may regard K as an extension of k(t) of degree 2, generated by y satisfying $y^2 = t^3 - 1$: evidently ramification occurs only at ∞ and at the cube roots of 1 in k, and so K/k has genus 1.

As in §1, we write $\beta_1 = X$, $\beta_2 = X - Y$, $\beta_3 = X + Y$, so that β_1, β_2, β_3 are elements of $\mathcal{O}$ satisfying

$$\beta_1\beta_2\beta_3 = 1 \quad \text{and} \quad -2\beta_1 + \beta_2 + \beta_3 = 0.$$

Now, if v is a finite valuation on K, then $v(\beta_i) = 0$ for $i = 1,2,3$. We wish to compute all the possibilities for the 3×3 matrix $\underline{A}$ with entries $v_i(\beta_j)$, $1 \leq i,j \leq 3$, where v_1, v_2, v_3 are the 3 infinite valuations on K. Applying the fundamental inequality (11) to the equation $-2\beta_1 + \beta_2 + \beta_3 = 0$, with V as the set of infinite valuations on K, we obtain

$$H(\beta_p/\beta_q) \leq 3 \qquad (1 \leq p,q \leq 3),$$

and hence

$$v_i(\beta_p) - v_i(\beta_q) \le 3 \qquad (1 \le i,p,q \le 3).$$

Further, since

$$v_i(\beta_1) + v_i(\beta_2) + v_i(\beta_3) = v_i(\beta_1\beta_2\beta_3) = 0 \qquad (1 \le i \le 3),$$

we obtain

$$|v_i(\beta_j)| \le 2 \qquad (1 \le i,j \le 3),$$

Thus there are only a finite number of possibilities for the matrix $\underline{A}$.

However, the number of possibilities may always be reduced substantially by further elementary considerations. For each $i = 1,2,3$, choose the indices p,q,r in such an order that $v_i(\beta_p) \le v_i(\beta_q) \le v_i(\beta_r)$. Since $\beta_2 + \beta_3 = 2\beta_1$, we have $v_i(\beta_p) \ge \min(v_i(\beta_q), v_i(\beta_r))$, and so $v_i(\beta_p) = v_i(\beta_q)$. This, together with the above, implies that $v_i(\beta_r) = -2v_i(\beta_p)$ and $v_i(\beta_p) = 0$ or -1. Furthermore, the sum formula (1) yields

$$\sum_{i=1}^{3} v_i(\beta_j) = 0 \qquad (1 \le j \le 3),$$

since $v(\beta_j) = 0$ for $v \nmid \infty$. We deduce that there are actually only seven possibilities for the matrix $\underline{A}$, namely the zero matrix, the matrix

$$\underline{B} = \begin{bmatrix} 2 & -1 & -1 \\ -1 & 2 & -1 \\ -1 & -1 & 2 \end{bmatrix},$$

and the five other matrices obtained by permuting the columns of $\underline{B}$. If $\underline{A}$ is the zero matrix, then $v(\beta_j) = 0$ for all the valuations on K, both finite and infinite, and so each β_j is an element of k. This possibility for $\underline{A}$ thus leads to the infinite family of solutions X, Y in k.

Now let us suppose that $\underline{A} = \underline{B}$. If we determine a non-zero element δ_1 in $\mathcal{O}$ such that $v_i(\delta_1) \ge v_i(\beta_1)$ for $i = 1,2,3$, then, as in the argument of Lemma 1, the sum formula (1) yields

$$0 = \sum_v v(\delta_1) \geq \sum_v v(\beta_1) = 0,$$

since $v(\delta_1) \geq 0$ if $v \nmid \infty$. Thus equality holds throughout, $v(\delta_1) = v(\beta_1)$ for all valuations v on K, and so β_1/δ_1 is an element of k. In order to determine such an element δ_1 we employ the construction in Lemma 1; rather than using the basis $1,y,y^2,y^3$ of K as there, we shall simplify the calculation by using the basis $1,y,t,yt$, which actually forms a $k[z]$-basis of $\mathcal{O}$. We may now express δ_1 uniquely in the form

$$\delta_1 = h_1 + h_2y + h_3t + h_4yt,$$

where h_1,h_2,h_3,h_4 are elements of $k[z]$. If τ runs over the embeddings corresponding to the infinite valuations on K, two for v_1 and one each for v_2 and v_3, then the four equations

$$\tau(\delta_1) = h_1 + h_2\tau(y) + h_3\tau(t) + h_4\tau(yt)$$

serve to express each of h_1,h_2,h_3,h_4 in terms of the $\tau(\delta_1)$. From these expressions and the inequalities $v_i(\delta_1) \geq v_i(\beta_1)$ we deduce that $\deg h_1 \leq 1$, $\deg h_2 \leq 1$, $\deg h_3 \leq 0$, $\deg h_4 \leq 0$, where we recall that $v(P) = -e_v \deg P$ for $v|\infty$ and P in $k[z]$. Thus there exist elements $a_1,\ldots,a_6$ in k, not all zero, such that

$$\delta_1 = a_1 + a_2z + a_3y + a_4yz + a_5t + a_6yt.$$

The conditions $v_2(\delta_1) \geq -1$, $v_3(\delta_1) \geq -1$ are satisfied, regardless of the values of $a_1,\ldots,a_6$, whilst the condition $v_1(\delta_1) \geq 2$ is equivalent, by consideration of the first Puiseux expansion above of y and t, to the linear equations

$$a_1 = a_3 = a_4 = a_6 = 0 \quad , \quad a_2 = a_5$$

Thus $\delta_1 = a_2(t+z) = -a_2/t$. Similarly we may determine the possibilities for β_2 and β_3 up to factors in k; in fact, if $\delta_1 = 1/t$, $\delta_2 = (y-\chi)/t$, $\delta_3 = (y+\chi)/t$, then each $\phi_j = \beta_j/\delta_j$, $j = 1,2,3$, is a non-zero element of k. Furthermore,

$$\phi_1\phi_2\phi_3 = 1 \qquad \text{and} \qquad -2\phi_1 + \phi_2(y-\chi) + \phi_3(y+\chi) = 0.$$

Differentiating the second equation we deduce that $\phi_2 + \phi_3 = 0$, since $y' \neq 0$; hence

$$\phi_2 = \phi_1 \chi \ , \ \phi_1^3 = 1 \ , \text{ and } \ X = \beta_1 = \phi_1/t \ , \ Y = \beta_1 - \beta_2 = -\phi_1 \chi y/t.$$

The five other possibilities for the matrix $\underline{A}$ may be dealt with similarly. Alternatively, it may be observed that the two equations $\beta_1\beta_2\beta_3 = 1$ and $-2\beta_1 + \beta_2 + \beta_3 = 0$ are symmetric in the 3 variables $-2\beta_1$, β_2, β_3. The five other possibilities for $\underline{A}$ merely correspond to the 5 permutations of these three variables. We conclude that the solutions X, Y in $\mathcal{O}$ of $X(X-Y)(X+Y) = 1$, other than X, Y in k, are

$$X = \phi/t, \quad Y = \phi\chi y/t \qquad \text{and} \qquad X = \phi(\chi y - 1)/2t, \quad Y = \pm\,\phi(\chi y + 3)/2t,$$

where χ, ϕ are any elements of k satisfying $\chi^2 = -1$ and $\phi^3 = 1$. Thus there are 18 isolated solutions and one infinite family.

CHAPTER III THE HYPERELLIPTIC EQUATION

1 *THE SOLUTION WHEN* L *IS FIXED*

In this chapter we shall solve completely the general hyperelliptic equation over an arbitrary algebraic function field K of characteristic zero. As for the Thue equation in Chapter II, it is the fundamental inequality, Lemma 2, which plays a crucial role in the analysis. However, in order that the fundamental inequality may be applied, we shall see that it is necessary to investigate the construction of division points on abelian varieties; this will be performed in §3. We shall in fact prove the following theorems; $\alpha_1,\ldots,\alpha_n$ denote $n(\geq 3)$ distinct elements of $\mathcal{O}$.

Theorem 4

All the solutions X, Y *in* $\mathcal{O}$ *of the hyperelliptic equation*

$$Y^2 = (X - \alpha_1)\ldots(X - \alpha_n) \qquad (14)$$

may be determined effectively.

Theorem 5

The equation (14) *has an infinite number of solutions in* $\mathcal{O}$ *if and only if there exists a substitution*

$$X = \alpha x + \beta \quad , \quad Y = \gamma y \qquad (\alpha,\beta,\gamma\in\mathcal{O}, \alpha\gamma\neq 0)$$

which transforms (14) *into* $y^2 = F(x)$, *where* F *is a polynomial with coefficients in the ground field* k. *Moreover, if such a substitution exists then there will be the infinite family derived from the solutions* x, y *in* k *of* $y^2 = F(x)$, *but otherwise* (14) *has only finitely many solutions* X, Y *in* $\mathcal{O}$.

Theorem 6

All the solutions X, Y *in* $\mathcal{O}$ *of* (14) *satisfy*

$$H(X) \leq 26H + 8g + 4(r-1);$$

here H *denotes the height of the polynomial on the right hand side of* (14), g *denotes the genus of* K/k *and* r *denotes the number of infinite valuations on* K.

The bound on the height of X in Theorem 6 evidently provides a bound on the height of Y. Furthermore, Theorems 4 and 6 also apply when $\alpha_1,\ldots,\alpha_n$ are no longer necessarily distinct, but at least three occur with odd multiplicity. Moreover, the bound on the height of X may be compared with that of Schmidt, namely $10^6(H+g+r)$ *[36]*.

For the proofs of the three theorems on the hyperelliptic equation we shall follow Siegel's original approach *[39]* and consider the field L generated over K by the square roots of $X-\alpha_1$, $X-\alpha_2$ and $X-\alpha_3$. In §3 we shall show that there are only a finite number of possibilities for L as X, Y run over the solutions in $\mathcal{O}$ of (14). In the classical case of algebraic numbers this result is an immediate consequence of the finiteness of the class group; here we shall show it may be deduced from a well known result concerning torsion points on abelian varieties. Moreover, by exhibiting an algorithm for the actual construction of such torsion points, we shall show in §3 that the possibilities for L are effectively determinable. It will suffice, therefore, for the proofs of Theorems 4 and 5 in the present section, to assume that $X-\alpha_1$, $X-\alpha_2$ and $X-\alpha_3$ are all squares in a *fixed* field L. The proof of Theorem 6, in §2, will be seen to result from a combination of the fundamental inequality (11) and the determination of a bound on the genus of L. Finally, in §4 we shall illustrate the complete algorithm by solving a particular elliptic equation in polynomials: as for the Thue equation no machine computation is necessary, a further reflection on the sharpness of the fundamental inequality.

In this section we shall assume that L is a given extension field of K; we wish to determine effectively all the solutions X, Y in $\mathcal{O}$ of (14) subject to the condition that there exist elements ξ_1, ξ_2 and ξ_3 in L with $\xi_i^2 = X-\alpha_i$, $i = 1,2,3$. We denote by $\mathcal{O}_L$ the ring of elements of L integral over $k[z]$, so that ξ_1, ξ_2 and ξ_3 are evidently elements of $\mathcal{O}_L$. Now let us define β_i, $\hat{\beta}_i$ in $\mathcal{O}_L$, $i = 1,2,3$, by $\beta_1 = \xi_2-\xi_3$, $\hat{\beta}_1 = \xi_2+\xi_3$, with β_2, $\hat{\beta}_2$, β_3 and $\hat{\beta}_3$ defined similarly by permutation of suffixes. Then $\beta_1\hat{\beta}_1 = \alpha_3-\alpha_2$, $\beta_2\hat{\beta}_2 = \alpha_1-\alpha_3$, $\beta_3\hat{\beta}_3 = \alpha_2-\alpha_1$, and

$$\beta_1 + \beta_2 + \beta_3 = 0 ; \qquad (15)$$

further, three similar equations may be obtained by conjugation, namely

$$\beta_1 + \hat{\beta}_2 - \hat{\beta}_3 = -\hat{\beta}_1 + \beta_2 + \hat{\beta}_3 = \hat{\beta}_1 - \hat{\beta}_2 + \beta_3 = 0. \qquad (16)$$

It is these four identities which form the crux of the analysis, both here and in the classical case of algebraic numbers, solved ineffectively by Siegel and effectively by Baker. In the classical case each β_i is factorised as $\delta_i\varepsilon_i^3$, where δ_i is an element of bounded height and ε_i is a unit; this is possible since β_i divides the fixed element $\mu = (\alpha_2-\alpha_3)(\alpha_3-\alpha_1)(\alpha_1-\alpha_2)$. The identity (15) is thereby transformed into a Thue equation in $\varepsilon_2/\varepsilon_1$ and $\varepsilon_3/\varepsilon_1$, namely

$$\delta_2(\varepsilon_2/\varepsilon_1)^3 + \delta_3(\varepsilon_3/\varepsilon_1)^3 = -\delta_1 ;$$

other Thue equations may be derived from (16). Using his bounds for the general Thue equation, Baker obtained *[6]* explicit bounds for the integer solutions of the hyperelliptic equation; these bounds are multiply exponential with respect to the original coefficients. In the case of function fields, a modification of this procedure, involving an extensive use of the Riemann-Roch theorem, is made by Schmidt *[36]*; he transforms (15) into an infinite set of Thue equations, whose coefficients nevertheless have bounded heights. Our fundamental inequality, however, enables us to solve the equation (15) directly, without reduction to a system of Thue equations as in previous work. It will be noted, moreover, that the bound thereby obtained in Theorem 6 varies only as a linear function of the height of the hyperelliptic equation, in contrast with the multiply exponential bounds for the classical case.

We now wish to establish Theorems 4 and 5, under the continuing assumption that L is fixed. We note as above that each of β_1, β_2 and β_3 divide in $\mathcal{O}_L$ the fixed element $\mu = (\alpha_2-\alpha_3)(\alpha_3-\alpha_1)(\alpha_1-\alpha_2)$: thus if w is any finite valuation on L such that $w(\mu) = 0$, then each $w(\beta_i) = 0$, and so $w(\hat{\beta}_i) = 0$ also. Hence Lemma 2 and its corollary may be applied to the equation $\beta_1+\beta_2+\beta_3 = 0$ and the set W of infinite valuations on L together with those finite valuations w for which $w(\mu)>0$. We deduce from the corollary that either β_1/β_2 is an element of k, or it belongs to some finite effectively determinable subset of $L\setminus k$. Let us now show that the ratio $\beta_1/\beta_2 = f$ determines X uniquely. For brevity we shall denote by κ the fixed element $(\alpha_1-\alpha_3)/(\alpha_2-\alpha_1)$ of K, so $(\beta_2\hat{\beta}_2)/(\beta_3\hat{\beta}_3) = \kappa$. From the equation $\beta_1+\beta_2+\beta_3 = 0$ we obtain $\beta_3/\beta_2 = -(1+f)$, and so $\hat{\beta}_2/\hat{\beta}_3 = -\kappa(1+f)$. The equation $\beta_1+\hat{\beta}_2-\hat{\beta}_3 = 0$ in (16) now yields $\beta_1/\hat{\beta}_3 = 1+\kappa+\kappa f$, and since $\beta_1/\beta_3 = -f/(1+f)$, we have $\beta_3/\hat{\beta}_3 = -(1+f)(1+\kappa+\kappa f)/f$, so f determines the ratio $\beta_3/\hat{\beta}_3$. But $\beta_3\hat{\beta}_3 = \alpha_2-\alpha_1$ and so

$$(X-\alpha_1)/(\alpha_2-\alpha_1) = \xi_1^2/\beta_3\hat{\beta}_3 = \tfrac{1}{4}(\beta_3+\hat{\beta}_3)^2/\beta_3\hat{\beta}_3, \qquad (17)$$

and thus f determines X as required.

We now distinguish between two types of hyperelliptic equation and prove Theorems 4 and 5 for both types separately. In the first case at least one of the ratios $\kappa_i = (\alpha_1-\alpha_i)/(\alpha_2-\alpha_1)$, $3\le i\le n$, is not an element of k. It follows immediately that no substitution of the type specified in Theorem 5 exists, and so we are required to prove that X has only finitely many possibilities which may be determined effectively. Rearranging $\alpha_3,\ldots,\alpha_n$ if necessary, it may be assumed that in fact $\kappa_3 = \kappa$ is not an element of k. Since the product of β_2/β_3 and $\hat{\beta}_2/\hat{\beta}_3$ is κ, we deduce that at least one of these two ratios is not an element of k. However, the corollary to Lemma 2 may be applied to each of the equations $\beta_1+\beta_2+\beta_3 = 0$ and $\beta_1+\hat{\beta}_2-\hat{\beta}_3 = 0$, from which we conclude that at least one of β_2/β_3 and $\hat{\beta}_2/\hat{\beta}_3$ belongs to a finite effectively determinable subset of L. Since each of these ratios determines f, which in turn determines X, we deduce that X has only finitely many effectively determinable possibilities in $\mathcal{O}$, as required. Hence Theorems 4 and 5 are established in this case.

We now assume that the alternative case holds, so that each κ_i, $1\le i\le n$, lies in the ground field k. As before, X is uniquely determined by $f = \beta_1/\beta_2$, which is either an element of k, or a member of a finite computable subset of $L\setminus k$. In this case we shall prove that any solution X, Y in $\mathcal{O}$ with Y non-zero such that f lies in k is derived from the solutions in k of $y^2 = \prod_{i=1}^{n}(x+\kappa_i)$ via a substitution of the form

$$X = \alpha x + \beta \;, \quad Y = \gamma y \qquad (\alpha,\beta,\gamma\in\mathcal{O},\ \alpha\gamma\neq 0)$$

as specified in Theorem 5; thus the proofs of Theorems 4 and 5 will be complete. For if f lies in k, then since $\kappa_3 = \kappa$ does also, we deduce that $\beta_3/\hat{\beta}_3$, equal to $-(1+f)(1+\kappa+\kappa f)/f$, and $x = (X-\alpha_1)/(\alpha_2-\alpha_1)$, given by (17), are also both elements of k. The hyperelliptic equation (14) may now be written as $Y^2 = (\alpha_2-\alpha_1)^n \prod_{i=1}^{n}(x+\kappa_i)$, in which the coefficients $\kappa_1,\ldots,\kappa_n$ lie in k. Assuming that Y is non-zero, this equation possesses a solution with x in k and Y in $\mathcal{O}$ if and only if $(\alpha_2-\alpha_1)^n$ is a square in $\mathcal{O}$. If this is the case, then writing $Y = \gamma y$, where $\gamma^2 = (\alpha_2-\alpha_1)^n$, transforms the above equation into $y^2 = \prod_{i=1}^{n}(x+\kappa_i)$, as required. We conclude that Theorems 4 and 5 are established in this case also.

We have now established Theorems 4 and 5 under the assumption that the field L is fixed. To complete the proof of Theorem 5 it suffices to show that L has only finitely many possibilities; the proof of Theorem 4 is completed by a demonstration that these possibilities may be determined effectively: both these aims will be accomplished in §3.

2 *BOUNDS*

This section will be devoted to a proof of Theorem 6, that for any solution X, Y in $\mathcal{O}$ of the hyperelliptic equation (14), we have

$$H(X) \leq 26H + 8g + 4(r-1),$$

where H is the height of the polynomial on the right hand side of (14), g is the genus of K/k and r is the number of infinite valuations on K. As discussed in §1, this bound may be compared favourably, both with that of Schmidt *[36]*, namely $10^6(H+g+r)$, and the bounds, obtained by Baker in the classical case of number fields, which are multiply exponential with respect to the maximum of the heights of the coefficients.

As in §1 we denote by $\mathcal{W}$ the set of valuations w on L such that $w|\infty$ or $w(\mu) > 0$. Furthermore, let us write for brevity $M = 2g_L-1+|\mathcal{W}|$, where g_L denotes the genus of L/k; since $|\mathcal{W}| \geq 1$ we have $M \geq 0$. We have already observed that, if w is any valuation outside $\mathcal{W}$, then $w(\beta_i) = w(\hat{\beta}_i) = 0$ for $i = 1,2,3$. Applying the fundamental inequality (11) to the equation $\beta_2+\beta_3+\beta_1 = 0$, we obtain

$$H_L(\beta_2/\beta_3) \leq M,$$

where $H_L(h)$ denotes the height of an element h in L, as defined in

Chapter I. Further, in view of the equations (16), M also serves as an upper bound for each of $H_L(\hat{\beta}_2/\beta_3)$, $H_L(\beta_2/\hat{\beta}_3)$ and $H_L(\hat{\beta}_2/\hat{\beta}_3)$. However, these four ratios, whose heights are bounded by M, provide a simple expression for X: since $\beta_2\hat{\beta}_2 = \alpha_1-\alpha_3$, $\beta_3\hat{\beta}_3 = \alpha_2-\alpha_1$ and $X-\alpha_1 = \xi_1^2 = \frac{1}{4}(\hat{\beta}_2-\beta_2)^2$, it follows that

$$2(2X-\alpha_1-\alpha_3)/(\alpha_2-\alpha_1) \quad = \quad (\hat{\beta}_2/\beta_3)(\hat{\beta}_2/\hat{\beta}_3) + (\beta_2/\beta_3)(\beta_2/\hat{\beta}_3).$$

Hence by (3) we have

$$H_L((2X-\alpha_1-\alpha_3)/(\alpha_2-\alpha_1)) \quad \le \quad 4M,$$

and so it will in fact suffice to bound the integer M.

As in Chapter I, §2, we denote by δ the degree of the extension L/K, and if w is a valuation on L, then w restricted to K has value group $\varepsilon_w\mathbb{Z}$ for some integer $\varepsilon_w > 0$, and $w|v$, where $w(f) = \varepsilon_w v(f)$ for f in K. Puiseux's theorem yields $\sum_{w|v} \varepsilon_w = \delta$ for each valuation v on K, and the genera of L and K are related by

$$(2g_L-2) - \delta(2g-2) \quad = \quad \sum_w (\varepsilon_w-1). \tag{7}$$

Let us denote by V the set of valuations v on K for which $v|\infty$ or $v(\mu) > 0$, so that W forms the set of valuations w on L such that $w|v$ for some v in V. Hence by (7) we obtain

$$M \le \delta(2g-1+|V|) + \sum_{w \notin W} (\varepsilon_w-1) \ ,$$

and so

$$M \le \delta(2g+r-1+|S|) \ ,$$

where S is the set of finite valuations v on K such that v lies in V or $\varepsilon_w > 1$ for some w on L such that $w|v$. It is therefore sufficient to establish the inequality $|S| \le 6H$. For if h lies in K then $H_L(h) = \delta H(h)$, and so

$$H((2X-\alpha_1-\alpha_3)/(\alpha_2-\alpha_1)) \quad \le \quad 4(2g+r-1+|S|).$$

Hence by (3) we have

$$H(X) \le 2H + 8g + 4(r-1) + 4|S| ,$$

since $\sum_{i=1}^{n} H(\alpha_i) = H$; Theorem 6 now follows if $|S| \le 6H$.

In order to establish the inequality $|S| \le 6H$, we show first that each v in S satisfies $v(\alpha_i-\alpha_j) > 0$ for some $j\neq i$, $1\le i\le 3$; we shall see that this fact is of crucial importance in the next section when we establish the range of possibilities for L. Suppose then that $v\nmid\infty$ satisfies $v(\alpha_i-\alpha_j) = 0$ for $1\le i<j\le n$, $1\le i\le 3$; we wish to prove that $\varepsilon_w = 1$ for each $w|v$. However, for each $i = 1,2,3$, we have $v(X-\alpha_i) = 0$ or $2v(Y)$; in either case the equation $X-\alpha_i = \xi_i^2$ is soluble for ξ_i in a Laurent series in powers of z_v, and so $\varepsilon_w = 1$ as required.

Let us now denote by S_i the set of finite valuations v on K such that $v(\alpha_i-\alpha_j) > 0$ for some $j\neq i$; in view of the last paragraph S is contained in the set $S_1\cup S_2\cup S_3$. Now

$$|S_i| \le \sum_{j\neq i}\ \sum_{v\nmid\infty} \max(0,v(\alpha_i-\alpha_j)) \le \sum_{j\neq i} H(\alpha_i-\alpha_j).$$

Since $\sum_{j=1}^{n} H(\alpha_j) = H$, the sum on the right is at most $(n-2)H(\alpha_i) + H$. Assuming as we may that $\alpha_1,\dots,\alpha_n$ are arranged in order of increasing height, it follows that $\sum_{i=1}^{3} H(\alpha_i) \le 3H/n$, from which we obtain $|S| \le 6H$ as required.

3 *THE DETERMINATION OF THE FIELD* L

In this section we shall complete the proofs of Theorems 4 and 5 by showing that there are only a finite number of effectively determinable possibilities for the field L as X, Y run over the solutions in $\mathcal{O}$ of the hyperelliptic equation (14), where L is generated over K by the square roots of $X-\alpha_1$, $X-\alpha_2$ and $X-\alpha_3$. Clearly it suffices to compute a finite set $f_1,\dots,f_m$ of non-zero elements of K such that, given any solution X, Y in $\mathcal{O}$ of (14), some ratio $(X-\alpha_1)/f_i$, $1\le i\le m$, is a perfect square in K. In the first part of this section we shall show that such a finite set $f_1,\dots,f_m$ exists: this result will be deduced from a familiar theorem concerning torsion points on abelian varieties. The second part of this section will be devoted to the production of an algorithm by which $f_1,\dots,f_m$ may be constructed. We first recall from §2 that if $\mathcal{U}$

denotes the set of valuations v on K such that $v|\infty$ or $v(\alpha_1-\alpha_i) > 0$ for some $2 \leq i \leq n$, then $v(X-\alpha_1)$ is an *even* integer for each v outside $\mathcal{U}$. We observe that $\mathcal{U}$ is a finite computable set; this fact forms the starting point for the proofs.

It will be necessary to employ the notions of a divisor on K and the divisor class group, which we recall briefly here (see, for example, *[11]*). A *divisor* $\mathfrak{a}$ consists of a set of integer components a_v, one for each valuation v on K, such that only finitely many a_v are non-zero. The degree $d(\mathfrak{a})$ of the divisor $\mathfrak{a}$ is then defined to be the sum $\sum_v a_v$. The set of all divisors on K forms an abelian group under addition by components, and within this group lies the subgroup of divisors of degree 0. If f is a non-zero element of K, then we define the *principal divisor* (f) associated with f to be the divisor with components v(f). The sum formula (1) states that all principal divisors have degree zero; furthermore, if f and h are both non-zero elements of K, then $v(fh) = v(f) + v(h)$ for each v, and so $(fh) = (f) + (h)$. Hence the principal divisors on K form a subgroup of the group of all divisors of degree zero; the quotient group J is termed the *divisor class group* of K. It is well known (see *[16]*) that J may be regarded as an abelian variety of dimension g, called the *Jacobian variety* of K. Furthermore, it is established in *[30, page 64]* that for any abelian variety J of dimension g, and any element $\mathfrak{B}$ of J, there are exactly 4^g solutions $\mathfrak{A}$ in J of the equation $2\mathfrak{A} = \mathfrak{B}$.

This last result yields the existence of a finite set $f_1,\ldots,f_m$. For, if X, Y forms a solution in $\mathcal{O}$ of (14), then $v(X-\alpha_1)$ is even for each v not in the finite set $\mathcal{U}$. Hence the principal divisor $(X-\alpha_1)$ is of the form $2\mathfrak{a}+\mathfrak{b}$, where $\mathfrak{a}$ and $\mathfrak{b}$ are divisors on K and $\mathfrak{b}$ has only finitely many possibilities; indeed, $\mathfrak{b}$ may be assumed to have components 0 or 1 for v in $\mathcal{U}$, with all components outside $\mathcal{U}$ being 0. Since $2\mathfrak{a}+\mathfrak{b}$ has degree zero, it follows that $d(\mathfrak{b}) = -2d(\mathfrak{a})$ and so, by adding a component $-d(\mathfrak{a})$ to $\mathfrak{a}$ and $-d(\mathfrak{b})$ to $\mathfrak{b}$ at some fixed infinite valuation v_0, it may be assumed that in fact both $\mathfrak{a}$ and $\mathfrak{b}$ have degree zero. Now $\mathfrak{b}$ has only finitely many possibilities which are effectively determinable, so we shall suppose throughout that $\mathfrak{b}$ is fixed. In the quotient group J we have $2[\mathfrak{a}] = -[\mathfrak{b}]$, where $[\mathfrak{a}]$ and $[\mathfrak{b}]$ denote the images in J of the divisors $\mathfrak{a}$ and $\mathfrak{b}$ respectively. However, we may now invoke the result *[30]* disclosed above, and deduce that this equation has exactly $4^g = p$ solutions in J, given by $[\mathfrak{a}_1],\ldots,[\mathfrak{a}_p]$, say. We may choose, for each $i=1,\ldots,p$, an element F_i in K such that $(F_i) = 2\mathfrak{a}_i + \mathfrak{b}$. It now follows that if $(F) = 2\mathfrak{a} + \mathfrak{b}$ for some divisor $\mathfrak{a}$, then there exists i, $1 \le i \le p$, with $[\mathfrak{a}] = [\mathfrak{a}_i]$, and so F/F_i is a square in K as required. Taking the union of the sets $F_1,\ldots,F_p$ over the finitely many possibilities for $\mathfrak{b}$, we achieve a finite set $f_1,\ldots,f_m$ with the desired property. We conclude that there are only a finite number of possibilities for the field L, and so the proof of Theorem 5 is complete.

Before we embark on the construction of the set $f_1,\ldots,f_m$, let us remark in passing that it is not necessary to invoke the algebraic-geometrical theory of general abelian varieties in order to establish the above result. For let us denote by $L(K^*)$ the set of non-zero f in K such that v(f) is even for all v, and by $(K^*)^2$ the set of all non-zero squares in K. Then it suffices for Theorem 5 to prove that the multiplicative group $L(K^*)/(K^*)^2$ is finite. Let us first consider the case when k is the field $\mathbb{C}$ of complex numbers. In this case the set of valuations on K may be identified with the points of a compact Riemann surface R of genus g, the elements of K corresponding to the meromorphic functions on R. By considering the pairing $(f,C) \mapsto \frac{1}{2\pi i}\int_C \frac{df}{f}$, where f is a non-zero element in K, and C is a closed curve on R not containing any pole or zero of f, it may be deduced that the group $L(K^*)/(K^*)^2$ is at most as large as the additive group $H_1(R)/2H_1(R)$; $H_1(R)$ denotes the first homology group of R (see [24]). Since the latter is isomorphic to $\mathbb{Z}^{2g}$, $H_1(R)/2H_1(R)$ has precisely 4^g elements, and so we conclude that $L(K^*)/(K^*)^2$ has at most 4^g elements, and so is finite as required. The result for an arbitrary algebraically closed field k of characteristic zero follows immediately from the familiar principle of Lefschetz [18].

In the second part of this section we shall prove that, given any divisor $\mathfrak{b}$ of degree zero, a set of representatives may be determined effectively for the p solutions of $2[\mathfrak{a}] = -[\mathfrak{b}]$ in J. This will complete the proof of Theorem 4, since, having constructed such divisors $\mathfrak{a}_1,\ldots,\mathfrak{a}_p$ then elements F_i such that $(F_i) = 2\mathfrak{a}_i + \mathfrak{b}$ may be determined effectively by Lemma 1. Thus a complete set $f_1,\ldots,f_m$ may be computed, such that, for each solution X, Y in $\mathcal{O}$ of (14), some ratio $(X-\alpha_1)/f_i$, $1 \le i \le m$, is a square in K. Hence all the possibilities for the field L may be determined effectively as required, and so the proof of Theorem 4 will be complete. The actual construction of the divisors $\mathfrak{a}_1,\ldots,\mathfrak{a}_p$, is achieved in three stages. First we shall choose for each divisor class in J a unique representative divisor, defined by certain properties (see below).

Secondly we shall show that these conditions on such a divisor $\mathfrak{a}$, together with those expressed by the equation $2[\mathfrak{a}] = -[\mathfrak{b}]$, are equivalent to a system of polynomial equations and inequalities over k in at most 2g unknowns. Now the resulting system of equations can possess at most *finitely* many solutions by construction, so the proof is completed by the third stage, Lemma 3, which states that whenever a system of polynomial equations and inequalities over k has only finitely many solutions, then these solutions may be determined effectively. It will then follow that the divisors $\mathfrak{a}_1, \ldots, \mathfrak{a}_p$ may be constructed, as required.

First then, we shall prove that for each divisor class in J there is a unique divisor $\mathfrak{a}$ in the class which has a single pole, of minimum order, at the infinite valuation v_0. We shall require here the Riemann-Roch theorem, which we recall briefly (see, for example, [11]). If $\mathfrak{a}$ and $\mathfrak{c}$ are two divisors on K, we say that $\mathfrak{a} \geq \mathfrak{c}$ if the components satisfy $a_v \geq c_v$ for each valuation v on K. It follows that for each divisor $\mathfrak{a}$, the set $V(\mathfrak{a})$ consisting of all elements f in K such that $f = 0$ or $(f) \geq -\mathfrak{a}$, forms a vector space over k; its dimension we denote by $\ell(\mathfrak{a})$. Since any principal divisor has degree 0, we have

$$\ell(\mathfrak{a}) = 0 \quad \text{if} \quad d(\mathfrak{a}) < 0.$$

Riemann's part of the Riemann-Roch theorem asserts that, on the other hand,

$$\ell(\mathfrak{a}) \geq 1 \quad \text{if} \quad d(\mathfrak{a}) \geq g ;$$

the Riemann-Roch theorem will be discussed in greater detail in Chapter IV. We shall denote by $\mathfrak{v}_0$ the divisor of degree 1 with a single non-zero component, at v_0. Let $\mathfrak{A}$ denote a divisor class in J; we wish to show that $\mathfrak{A}$ contains a unique divisor $\mathfrak{a}$ such that $\mathfrak{a} \geqslant -s\mathfrak{v}_0$ for some minimal integer s. We first select any divisor $\mathfrak{c}$ in $\mathfrak{A}$; then by Riemann's theorem we obtain $\ell(\mathfrak{c}+n\mathfrak{v}_0) \geq 1$ if $n \geq g$ since $d(\mathfrak{c}) = 0$. Now the condition that an element of $V(\mathfrak{c}+n\mathfrak{v}_0)$ lies in the subspace $V(\mathfrak{c}+n\mathfrak{v}_0-\mathfrak{v}_0)$ is equivalent to a single linear equation, and so the subspace has codimension 0 or 1. Hence there is a unique integer s, with $0 \leq s \leq g$, such that $\ell(\mathfrak{c}+s\mathfrak{v}_0) = 1$ and $\ell(\mathfrak{c}+s\mathfrak{v}_0-\mathfrak{v}_0) = 0$. If F is any non-zero element of $V(\mathfrak{c}+s\mathfrak{v}_0)$, then we choose for our unique representative of $\mathfrak{A}$ the divisor $\mathfrak{c} = \mathfrak{c}+(F)$: $\mathfrak{a}$ has a single pole, at v_0, of order s. Now $\mathfrak{a}$ is unique, for if $\mathfrak{a}'$ is another divisor in

$\mathfrak{A}$ such that $\mathfrak{a} \geqslant -s\mathfrak{v}_0$, then $\mathfrak{a}'-\mathfrak{a}$ is principal, equal to (h) say, and then $(h)+s\mathfrak{v}_0+\mathfrak{c} > 0$. Thus h lies in the space $V(\mathfrak{c}+s\mathfrak{v}_0)$ of dimension 1, so h/F is an element of k and $\mathfrak{a}' = \mathfrak{a}$ as required.

We have thus shown that each divisor class $\mathfrak{A}$ in J admits a unique representative $\mathfrak{a}$ such that $\mathfrak{a} \geqslant -s\mathfrak{v}_0$, where s is specified by the two equations $\ell(\mathfrak{c}+s\mathfrak{v}_0) = 1$, $\ell(\mathfrak{c}+s\mathfrak{v}_0-\mathfrak{v}_0) = 0$, $\mathfrak{c}$ being any element of $\mathfrak{A}$. We now wish to impose the further condition that $2[\mathfrak{a}] = -[\mathfrak{b}]$; that is, $2\mathfrak{a}+\mathfrak{b}$ is principal. Since $\ell(\mathfrak{d}+(h)) = \ell(\mathfrak{d})$ for any principal divisor (h) and any divisor $\mathfrak{d}$, the conditions that $2\mathfrak{a}+\mathfrak{b}$ is principal and $\ell(\mathfrak{c}+s\mathfrak{v}_0-\mathfrak{v}_0) = 0$ are equivalent to the two equations $\ell(-\mathfrak{b}-2\mathfrak{a}) = 1$ and $\ell(s\mathfrak{v}_0-\mathfrak{v}_0-\mathfrak{b}-\mathfrak{a}) = 0$. Furthermore, the condition $\ell(\mathfrak{c}+s\mathfrak{v}_0) = 1$ is now automatically satisfied, since $\mathfrak{a}$ lies in the same class as $\mathfrak{c}$ and $\mathfrak{a} \geq -s\mathfrak{v}_0$. We may summarise the conditions on the divisor $\mathfrak{a}$ as

$$d(\mathfrak{a}) = 0, \quad \mathfrak{a} \geq -s\mathfrak{v}_0, \quad \ell(s\mathfrak{v}_0-\mathfrak{v}_0-\mathfrak{b}-\mathfrak{a}) = 0, \quad \ell(-\mathfrak{b}-2\mathfrak{a}) = 1, \tag{18}$$

where s is some integer, $0 \leq s \leq g$. Our object is to determine effectively the p solutions $\mathfrak{a}$ of (18). We proceed to show that, for each s, the conditions in (18) may be expressed as a system of polynomial equations and inequalities over k in at most 2s unknowns. Since we know that this system may possess only finitely many solutions, it follows by Lemma 3, proved below, that the system may be solved effectively; the proof of Theorem 4 will be completed thereby.

The case s = 0 may be disposed of immediately: (18) then implies that $\mathfrak{a}$ is the zero divisor and $\mathfrak{b}$ is principal. It may be determined using Lemma 1 whether $\mathfrak{b}$ is principal and, if so, an element f such that $(f) = \mathfrak{b}$ may be computed. Such an element would then form one of the required set $f_1,\dots,f_m$. Henceforth we shall suppose that $1 \leq s \leq g$.

Let us first assume that the components of $\mathfrak{a}$ are equal to $-s$ at v_0, 1 at $v_1,\dots,v_s$, and 0 otherwise, so that $d(\mathfrak{a}) = 0$ and $\mathfrak{a} \geq -s\mathfrak{v}_0$; we wish to determine the valuations $v_1,\dots,v_s$. As in Chapter I we denote by P the minimal polynomial of the primitive element y in $\mathcal{O}$ over $k[z]$ and by D the discriminant of P. We shall further assume that each v_i, $1 \leq i \leq s$, is a finite valuation not in $\mathcal{U}$ such that $v_i(D) = 0$. The exceptional cases, when $\mathfrak{a}$ has a component larger than 1, or when some v_i is either in $\mathcal{U}$ or satisfies $v_i(D) > 0$, will be dealt with in the next paragraph. Now each valuation v_i is uniquely specified by the elements y_i and z_i in k such that $v_i(y-y_i) > 0$ and $v_i(z-z_i) > 0$; it is the 2s unknowns $y_1,\dots,y_s$,

$z_1,\ldots,z_s$ which are determined by a system of polynomial equations and inequalities, which we now exhibit. We recall from Lemma 1 that, whenever f lies in $\mathcal{O}$, then Df may be expressed as a polynomial in y and z, with coefficients in k. The assumptions on $v_1,\ldots,v_s$ may now be expressed as

$$P(y_i,z_i) = 0 \quad , \quad D\theta(y_i,z_i) \neq 0 \qquad (1\leq i\leq s),$$

where $\theta = \prod_{j=2}^{n}(\alpha_1-\alpha_j)$. Furthermore, the valuations $v_1,\ldots,v_s$ are distinct, and so

$$y_i \neq y_j \quad \text{or} \quad z_i \neq z_j \qquad (1\leq i<j\leq s).$$

It remains to express the two equations in (18) as polynomial equations and inequalities in $y_1,\ldots,z_s$. Let $H_1,\ldots,H_t$ denote a basis of the vector space $V(2s\mathfrak{v}_0-\mathfrak{b})$; such a basis may be computed using the same method as in Lemma 1. Since $\mathfrak{b}$ has only one pole, at the infinite valuation v_0, it follows that each H_i lies in $\mathcal{O}$, and so each DH_i is a polynomial in y and z. Furthermore, $y'\frac{\partial P}{\partial X}(y,z) = -\frac{\partial P}{\partial z}(y,z)$ and since $D = N(\frac{\partial P}{\partial X}(y,z))$, it follows that Dy' is a polynomial in y and z; we conclude that whenever f lies in $\mathcal{O}$, then D^2f' is a polynomial in y and z: we shall apply this result with $f = H_i$ for each i. If we denote by f(v) the constant coefficient of the Puiseux expansion of f at v, then f lies in $V(-\mathfrak{b}-2\mathfrak{a})$ if and only if $f = \sum_{i=1}^{t}\pi_iH_i$ for some $\pi_1,\ldots,\pi_t$ in k such that $f(v_j) = f'(v_j) = 0$, so $\sum_{i=1}^{t}\pi_iH_i(v_j) = \sum_{i=1}^{t}\pi_iH_i'(v_j) = 0$, for $1\leq j\leq s$. Since the polynomial D does not vanish at any v_j, these conditions on $\pi_1,\ldots,\pi_t$ are equivalent to $\sum_{i=1}^{t}\pi_i(DH_i)(y_j,z_j) = \sum_{i=1}^{t}\pi_i(D^2H_i')(y_j,z_j) = 0$ for $1\leq j\leq s$. The requirement that $\ell(-\mathfrak{b}-2\mathfrak{a}) = 1$ is therefore equivalent to the condition that the $t\times 2s$ matrix with entries $DH_i(y_j,z_j)$ and $D^2H_i'(y_j,z_j)$, $1\leq i\leq t$, $1\leq j\leq s$, has rank t-1. Since each entry in this matrix is a polynomial in y_j and z_j, the condition $\ell(-\mathfrak{b}-2\mathfrak{a}) = 1$ may be expressed as a system of polynomial equations and inequalities in $y_1,\ldots,z_s$. Similarly the condition $\ell(s\mathfrak{v}_0-\mathfrak{v}_0-\mathfrak{b}-\mathfrak{a}) = 0$ may be so expressed, and so the construction is complete in this case.

It remains to deal with a general divisor satisfying (18). We may suppose that has a component -s at v_0, components $s_1,\ldots,s_q$ at valuations $v_1,\ldots,v_q$ respectively, with all other non-zero components occurring at valuations v either in $\mathcal{U}$ or such that $v(D) > 0$. The sum of the positive components of $\mathfrak{a}$ is s, and so there are only a finite number of possibilities for the integers $s_1,\ldots,s_q$, and for the other non-zero components. Hence we may suppose that all these are fixed, and it only remains to determine the range of possibilities for the valuations $v_1,\ldots,v_q$. As before, our object is to express the conditions on $v_1,\ldots,v_q$ as a system of polynomial equations and inequalities in the 2q unknowns $y_1,\ldots,y_q,z_1,\ldots,z_q$. The conditions on $\mathfrak{a}$ in (18) are again equivalent to a determination of the ranks of two matrices, the first, corresponding to $\ell(-\mathfrak{b}-2\mathfrak{a}) = 1$, having entries of the form $D^{r_j+1}f^{(r_j)}(y_j,z_j)$, $0\le r_j\le 2s_j$, $1\le j\le q$, and the second, corresponding to $\ell(s\mathfrak{d}_0-\mathfrak{d}_0-\mathfrak{b}-\mathfrak{a}) = 0$ having entries of the same form but with $0\le r_j<s_j$, $1\le j\le q$; $f^{(n)}$ denotes the n-th derivative of f with respect to z. Now each entry in these two matrices is a polynomial in y_j and z_j, and so the conditions on the ranks are again equivalent to a system of polynomial equations and inequalities, as required.

We have now reduced the problem of constructing the divisors, and so all the possibilities for the field L, to that of solving a system of polynomial equations and inequalities over k. This we now provide an algorithm for the accomplishment thereof, thus completing the proof of Theorem 4.

Lemma 3

Let $F_1,\ldots,F_m,G_1,\ldots,G_n$ *denote polynomials in the* p *variables* $x_1,\ldots,x_p$ *with coefficients in* k. *Suppose that the system*

$$F_i(x_1,\ldots,x_p) = 0\ ,\ G_j(x_1,\ldots,x_p) \neq 0 \qquad (1\le i\le m, 1\le j\le n)$$

of equations and inequalities possesses only finitely many solutions $x_1,\ldots,x_p$ *in* k. *Then all the solutions may be determined effectively.*

Proof. (See [46]). If we introduce a new variable x_0 together with a new equation $F_0(x_0,\ldots,x_p) = 0$, where

$$F_0(x_0,\ldots,x_p) = x_0\prod_{j=1}^{n}G_j(x_1,\ldots,x_p) - 1,$$

then the solutions $x_1,\ldots,x_p$ of the original system are in exact correspondence with the solutions $x_0,\ldots,x_p$ of the system $F_0=\ldots=F_m=0$ of equations alone. It thus suffices to assume, as we shall, that the system to be solved consists of equations alone. The proof now proceeds by induction on the number of variables, p. If p=1, then at least one polynomial F_i is not identically zero. Since k is presented explicitly, all the zeros of F_i may be determined effectively, and each may be tested as a solution of the complete system. Hence the complete set of solutions of the system may be computed.

Let us now assume that $p \geq 2$, and that any system of equations in p-1 unknowns, which has only finitely many solutions in k, may be solved effectively. We wish to eliminate one of the variables $x_1,\ldots,x_p$ from the system, in order to apply this inductive hypothesis. We may suppose that the polynomial F_1 involves x_1 and has total degree Δ, say, where the total degree of a polynomial is the largest of the total degrees $\sum_{i=1}^{p} \Delta_i$ of the monomials $cx_1^{\Delta_1}\ldots x_p^{\Delta_p}$ occurring. By making the substitution $x_1 = c_1y_1,\ x_2 = y_2+c_2y_1,\ldots,x_p = y_p+c_py_1$, for some suitable choice of $c_1,\ldots,c_p$ in k, it may be assumed that F_1 contains a non-zero term cx_1^{Δ}. Now, if $x_2,\ldots,x_p$ are fixed, we observe that the polynomials $F_1,\ldots,F_m$ have a common zero x_1 if and only if the two polynomials F_1 and $\sum_{i=2}^{m} a_iF_i$ have a common zero for each choice of $a_2,\ldots,a_m$ in k. For since the polynomial F_1 involves cx_1^{Δ}, it may only have a finite number of zeros x_1 for each $x_2,\ldots,x_p$, and so one of these would then have to be a common zero of $F_2,\ldots,F_m$ as required. Furthermore, since the leading coefficient of F_1 with respect to x_1 is non-zero, the two polynomials F_1 and $\sum_{i=2}^{m} a_iF_i$ have a common zero x_1 if and only if the resultant $R_{x_1}(F_1, \sum_{i=2}^{m} a_iF_i)$ vanishes. However, this resultant is a polynomial in $x_2,\ldots,x_p,a_2,\ldots,a_p$, and so may be expressed in the form

$$\sum_{r_2,\ldots,r_m} \ldots \sum a_2^{r_2}\ldots a_m^{r_m}\, H_{\underline{r}}(x_2,\ldots,x_p),$$

where $\underline{r} = (r_2,\ldots,r_m)$, and each $H_{\underline{r}}$ is a polynomial in $x_2,\ldots,x_p$. We conclude that the polynomials $F_1,\ldots,F_m$ have a common zero $x_1,\ldots,x_p$ if and only if the polynomials $H_{\underline{r}}$ have a common zero $x_2,\ldots,x_p$ for all choices of $\underline{r}$. Since by construction this latter system possesses only finitely many solutions in k, all the solutions may be determined effectively by the inductive hypothesis. For each of the possibilities for $x_2,\ldots,x_p$, the common zeros of $F_1,\ldots,F_m$ may be determined as in the case p=1 above. Hence all the solutions of the original system may be determined effectively, as required.

The proof of Theorem 4 is now complete, and we conclude this section by making two remarks on the construction described in this section. First, the construction may be considerably simplified in practice by the following expedient. After constructing the solutions $\mathfrak{a}_i$, $1\le i\le p$, of the equations (18) with $\mathfrak{b} = 0$, it suffices to construct just one solution $\mathfrak{a}$ for the equations (18) for each of the finitely many possibilities for $\mathfrak{b}$. A set of representatives for the classes $\mathfrak{A}$ satisfying $2\mathfrak{A} = -[\mathfrak{b}]$ is then provided by the divisors $\mathfrak{a}+\mathfrak{a}_i$, $1\le i\le p$. If the elements of K corresponding to the divisors $\mathfrak{a}_i$ are F_i, $1\le i\le p$, and that corresponding to $\mathfrak{a}$ is F, so $(F) = 2\mathfrak{a}+\mathfrak{b}$ then the element of K corresponding to the divisor $\mathfrak{a}+\mathfrak{a}_i$ is FF_i. Furthermore, if $\mathfrak{a}$ and $\mathfrak{a}'$ are divisors such that $2\mathfrak{a}+\mathfrak{b}$ and $2\mathfrak{a}'+\mathfrak{b}'$ are principal, then $2(\mathfrak{a}+\mathfrak{a}')+\mathfrak{b}+\mathfrak{b}'$ is principal. It is therefore in fact sufficient to determine all the solutions of (18) with $\mathfrak{b} = 0$, and to determine just one solution for each $\mathfrak{b}$ which has just two non-zero components, -1 at v_o, and 1 at some valuation v in $\mathcal{U}$. This simplification considerably shortens the calculation.

Secondly, it is possible in principle to deduce Theorem 4 directly from Theorems 5, 6 and Lemma 3, thereby avoiding the actual construction of the divisors $\mathfrak{a}_1,\ldots,\mathfrak{a}_p$ which occupied the major part of this section. If $h_1,\ldots,h_d$ denotes a $k[z]$-basis of $\mathcal{O}$, then we may express any solution X, Y in $\mathcal{O}$ of the hyperelliptic equation (14) in the form

$$X = \sum_{j=0}^{J}\sum_{i=1}^{d} a_{ij}z^j h_i \qquad Y = \sum_{j=0}^{J}\sum_{i=1}^{d} b_{ij}z^j h_i ,$$

where the coefficients a_{ij} and b_{ij} lie in k. In view of the bound in Theorem 6, J is bounded. Furthermore, the hyperelliptic equation (14) is equivalent to a system of polynomial equations in the unknowns a_{ij}, b_{ij}. We may determine, using Theorem 5, whether these equations possess only finitely many solutions. If so, then, by virtue of Lemma 3, all the solutions may be determined effectively. If not, then the exceptional solutions outside the infinite family may be determined by imposing the extra inequality $((X-\alpha_1)/(\alpha_2-\alpha_1))' \ne 0$; the resulting system of equations and inequalities in the a_{ij}, b_{ij} may then be solved effectively. We have thus proved Theorem 4 by a considerably shorter method. However, this brevity is in fact deceptive since this algorithm is far less efficient than that described in §1 and the central part of this section. The complexity of the algorithm in Lemma 3 varies exponentially with the number of variables. In our earlier argument we have endeavoured to keep the number of variables as small as possible: at most $2g$. If the method just advanced were adopted, the number of variables would be $2d(J+1)$, which may be extremely large, and in any case depends on the height H of the equation (14). Even equations of comparatively small height would be incapable of solution in a reasonable time by machine computation using the second method, while readily within the scope of the first. For instance, the example in the next section would require the solution of some 157 equations in 132 unknowns using this second method, a most formidable task. We shall solve the equation using the first method, without recourse even to machine computation.

4 *AN EXAMPLE*

In this section we shall determine all the solutions X, Y in $k[z]$ of the elliptic equation

$$Y^2 = X(X - z)(X - 2z + 1).$$

This demonstration will provide a simple illustration of the general method of analysis described in §§1, 3. It will be seen that an exactly similar argument would apply to any equation of the form $Y^2 = G(X)$, where G has at least three zeros in $k(z)$ with odd multiplicities; the number of steps involved in the calculation varies exponentially with $H(G)$, that is, the degree of G in z.

We shall assume throughout that Y is non-zero. First we shall determine all the possibilities for the field L, generated over $k(z)$ by the square roots of X, $X-z$ and $X-2z+1$. If a is any finite point of k such that $\mathrm{ord}_a X > 0$, then X has a zero at a, and so either $a = 0, \frac{1}{2}$, or $\mathrm{ord}_a X = 2\,\mathrm{ord}_a Y$. We deduce that $\mathrm{ord}_a X$ is even for all $a \neq 0, \frac{1}{2}$, and so one of X, X/z, $X/(2z-1)$, $X/z(2z-1)$ is a square in $k[k]$. Similarly one of $X-z$, $(X-z)/z$, $(X-z)/(z-1)$, $(X-z)/z(z-1)$ and one of $X-2z+1$, $(X-2z+1)/(z-1)$, $(X-2z+1)/(2z-1)$, $(X-2z+1)/(z-1)(2z-1)$ are squares in $k[z]$. We conclude that L has at most 15 possibilities.

Let us first suppose that $X = h_1^2 z(2z-1)$ for some h_1 in $k[z]$. From degree parity arguments it follows that there exist h_2 and h_3 in $k[z]$ such that $X-z = h_2^2 z(z-1)$ and $X-2z+1 = h_3^2(2z-1)(z-1)$. In this case L is an extension of k(z) of degree 4, generated by the square roots of z(2z-1) and z(z-1). The only valuations v on L which are ramified are those $v|a$ with $a = 0$, ½ or 1, and all these have ramification index 2; it follows from the genus formula (7) that L/k has genus 0. We wish to determine, as in §1, all the possibilities for β_1, β_2 and β_3, where

$$\beta_1 = h_2\sqrt{z(z-1)} - h_3\sqrt{(2z-1)(z-1)},$$

$$\beta_2 = h_3\sqrt{(2z-1)(z-1)} - h_1\sqrt{z(2z-1)},$$

$$\beta_3 = h_1\sqrt{z(2z-1)} - h_2\sqrt{z(z-1)};$$

$\hat{\beta}_1$, $\hat{\beta}_2$, $\hat{\beta}_3$ are then given by

$$\hat{\beta}_1 = h_2\sqrt{z(z-1)} + h_3\sqrt{(2z-1)(z-1)},$$

$$\hat{\beta}_2 = h_3\sqrt{(2z-1)(z-1)} + h_1\sqrt{z(2z-1)},$$

$$\hat{\beta}_3 = h_1\sqrt{z(2z-1)} + h_2\sqrt{z(z-1)}.$$

Since $\beta_1\hat{\beta}_1 = z-1$ and h_2, h_3 lie in $k[z]$, we deduce that if v is a finite valuation on L, then $v(\beta_1) = v(\hat{\beta}_1) = 0$ or 1, according as $v(z-1) = 0$ or $v(z-1) > 0$. Similar deductions may be made for $v(\beta_2)$ and $v(\beta_3)$ at each finite valuation on L, so it only remains to investigate the values of β_1, β_2 and β_3 at the infinite valuations. Let us denote the four infinite valuations on L by v_1, v_2, v_3 and v_4; we wish to determine all the possibilities for the 4×3 matrix $\underline{B}$ with entries $v_i(\beta_j)$. Applying the fundamental inequality (11) to the equation $\beta_1+\beta_2+\beta_3 = 0$ with V as the set of 4 infinite valuations together with the 6 ramified valuations, we obtain $H(\beta_p/\beta_q) \le 8$ for $p \neq q$. Since $H(f) = \sum_v \max(0,v(f))$ for f non-zero, we deduce that

$$\sum_{i=1}^{4} \max(0,v_i(\beta_p)-v_i(\beta_q)) \le 6 \qquad (1\le p\neq q\le 3), \quad (19)$$

and so $|v_i(\beta_p)-v_i(\beta_q)| \le 6$. Furthermore, $v_i(\beta_j)+v_i(\hat{\beta}_j) = 1$ for each i, j, so applying the fundamental inequality (11) to each of the equations in (16), we obtain $|v_i(\beta_p)+v_i(\beta_q)| \le 7$, and so $|v_i(\beta_j)| \le 6$ for each i, j. Hence there are only a finite number of possibilities for the matrix $\underline{B}$. However, just as for the example on the Thue equation solved in Chapter II, §3, the number of possibilities may always be reduced substantially by some further elementary considerations. For each $i = 1,2,3,4$, let us choose indices p,q,r such that $v_i(\beta_p) \le v_i(\beta_q) \le v_i(\beta_r)$; it follows from $\beta_p + \beta_q + \beta_r = 0$ that $v_i(\beta_p) \ge \min(v_i(\beta_q), v_i(\beta_r))$ and so $v_i(\beta_p) = v_i(\beta_q)$. Applying this argument to the equations (16), we deduce that either $v_i(\beta_p) = v_i(\beta_q) = v_i(\beta_r) \ge 0$, or $v_i(\beta_p) = v_i(\beta_q) = -v_i(\beta_r) - 1 \le -1$. Furthermore, since $\sum_{v \nmid \infty} v(\beta_j) = 2$ for each j, the sum formula (1) yields $\sum_{i=1}^{4} v_i(\beta_j) = -2$. It now follows that the only possibilities for the matrix $\underline{B}$ are the matrices

$$\underline{T}(n) = \begin{bmatrix} n & n & n \\ n & m & m \\ m & n & m \\ m & m & n \end{bmatrix} \qquad (m+n=-1, n\ge 0),$$

or one of the matrices obtained from $\underline{T}(n)$ by permuting the rows. Taking into account the choice made in the labelling of the infinite valuations on L, and the fact that h_1, h_2 and h_3 lie in $k[z]$, we deduce that in fact only the matrices $\underline{T}(n)$ need be considered. In view of the inequality (19) above, we have $n \le 2$. Now $\deg h_3 = -v_i(\hat{\beta}_1-\beta_1)-1 = n$, so X has degree 2n+2. If n=0, then h_1, h_2 and h_3 are elements of k, and from the conditions $v_1(\beta_1) = v_1(\beta_2) = 0$ we obtain the linear equations $h_2-\sqrt{2}h_3 = h_3-h_1 = 0$; from (16) we then obtain $h_3^2 = 1$, and this yields the solutions $X = z(2z-1)$, $Y = \pm\sqrt{2}\, z(z-1)(2z-1)$. If n=1, then $h_3 = a+bz$ for some a, b in k with b non-zero. From the condition $v_1(\beta_1) = 1$ we obtain $8a+b = 0$, whilst from $v_1(\beta_2) = 1$ we obtain $4a+b = 0$, and these are incompatible with $b\ne 0$. Similarly we may rule out the case n=2: then $h_3 = a+bz+cz^2$ with a, b, c in k, c non-zero; from the conditions $v_1(\beta_1) = v_1(\beta_2) = 2$ we obtain $8a+2b+c = 16a+8b+5c = 32a+4b+c = 64a+16b+5c = 0$, and these are incompatible with $c\ne 0$.

Now let us assume that one of X, X/z, X/(2z-1) is a square in $k[z]$. In the first case, since Y^2/X is then a square in $k[z]$, T^2 say, we have

$$(X - \frac{3z-1}{2} + T)(X - \frac{3z-1}{2} - T) = \frac{(z-1)^2}{4}. \tag{20}$$

Both factors on the left are polynomials, so X has degree at most 2. If X lies in k, then, since $(z-X)(2z-1-X)$ is a square, $X = 1$. If X has degree 2, then, for some ψ in k, the two factors on the left of (20) are equal to $\frac{1}{4}\psi$ and $(z-1)^2/\psi$; since X is a square we deduce $\psi = 1$ and $X = \frac{1}{2}(z+\frac{1}{2})^2$. In the other cases, when X/z or $X(2z-1)$ is a square in $k[z]$, then (20) holds with T in $\mathcal{O}_K$, where $K = k(\sqrt{z})$ or $k(\sqrt{2z-1})$ respectively. In both cases K has only one infinite valuation, and so we deduce from (20) that X has degree at most 2. Thus if X/z is a square, then $X = \psi z$ for some ψ in k, whence $\psi = 2$, whilst if $X/(2z-1)$ is a square, then $X = \psi(2z-1)$ for some ψ in k, whence $\psi = \frac{1}{2}$.

We conclude that the solutions X, Y in $k[z]$ of $Y^2 = X(X-z)(X-2z+1)$ are $X = 0, z$ or $2z-1$, $Z = 0$; $X = z(2z-1)$, $Z = z(z-1)(2z-1)$; $X = 1$, $Z = z-1$; $X = \frac{1}{2}(z+\frac{1}{2})^2$, $Z = \frac{1}{4}(z-\frac{1}{2})(z+\frac{1}{2})(z-\frac{3}{2})$; $X = 2z$, $Z = z$; $X = z-\frac{1}{2}$, $Z = \frac{1}{2}(z-\frac{1}{2})$; in each case $Y = \pm\sqrt{2}\,Z$. Hence there are exactly 13 solutions.

CHAPTER IV EQUATIONS OF SMALL GENUS

1 *INTRODUCTION*

This chapter will be devoted to the construction of algorithms whereby may be determined the complete set of integral solutions of all equations of genera 0 and 1 over an arbitrary algebraic function field K. As usual *[40]* we shall assume that in the former case the curve associated with the equation possesses at least three infinite valuations. As for the Thue and hyperelliptic equations already solved, in each case we shall establish a simple criterion for the equation to possess an infinity of solutions in $\mathcal{O}$. In the next chapter a constructive examination of the algorithms will be used to establish explicit bounds on the heights of the solutions. The chief ingredient in those proofs is a direct recursive technique for determining the coefficients in a Puiseux series (see Lemma 5 and *[10]*). The bounds obtained are linear functions of the height of the original equation and thus are not exponential, as are the bounds established by Baker and Coates *[8]* in the case of algebraic numbers. This provides further evidence of the strength and power of our fundamental inequality which again plays the crucial role in the analysis. These results on the heights of solutions may be viewed in another way, as a complement to the celebrated theorem of Manin and Grauert, which had proved the analogue for function fields of Faltings' recent result. We recall that, as a consequence of this theorem, the heights of all solutions in K, not just those in $\mathcal{O}$, of any equation of genus 2 or more, are bounded. We further recall that Parsin has succeeded in deriving explicit bounds on the heights, thereby enabling the solutions in K to be determined, at least in principle, of any equation of genus 2 or more. However, in the case of genera 0 or 1 the algorithms derived here are efficient enough to enable equations to be solved directly. Returning to the analogous case of number fields, a determination of effective bounds on the solutions of equations of arbitrary genus remains an important unsolved problem: even the case of integral solutions of the general equation of genus 2 would represent a significant advance, when the boundedness follows from Siegel's famous theorem [40].

We shall first indulge in some preliminary discursion on the analogy between equations and function fields, and the geometry of curves including the Riemann-Roch theorem. As in previous chapters K will denote a finite extension of the rational function field k(z). We shall be concerned with the determination of all the solutions X, Y in $\mathcal{O}$ of an equation F(X,Y) = 0, where F denotes an absolutely irreducible polynomial with coefficients in K, that is F is irreducible over the algebraic closure of K. Let us denote such an algebraic closure by $\mathcal{k}$; like k, $\mathcal{k}$ is an algebraically closed field of characteristic zero, and a great deal of work may be eliminated by exploiting this similarity: to emphasise it we shall employ script letters to denote the constructs for $\mathcal{k}$ corresponding to those made for k. Furthermore, there exists an embedding σ of $\mathcal{k}$ in the algebraically closed field of all Puiseux series in fractional powers of 1/z, and this assists in enabling $\mathcal{k}$ to be presented explicitly. Each element ϕ of $\mathcal{k}$ may be specified uniquely by its minimal polynomial Φ over k(z) and a finite number of terms in the Puiseux series $\sigma(\phi)$, sufficient to distinguish ϕ from the other zeros of Φ. Since we may compute any further coefficients of the series $\sigma(\phi)$, and determine effectively whether a polynomial is irreducible over k(z), this description provides an explicit presentation of $\mathcal{k}$ as required. We may define on the rational function field $\mathcal{k}$(X) the valuations ord and, for each a in $\mathcal{k}$, ord_a; each element f of $\mathcal{k}$(X) may be expanded as a Laurent series over $\mathcal{k}$ in powers of either 1/X or X-a, and ord(f), ord_a(f) denote the orders of vanishing of the respective series. Now F is irreducible over $\mathcal{k}$, so we may adjoin to $\mathcal{k}$(X) a root Y of F(X,Y) = 0 and thereby obtain a field $\mathcal{K}$, of degree $d = \deg_Y F$ over $\mathcal{k}$(X). In view of Puiseux's theorem we may construct valuations on $\mathcal{K}$, just as in Chapter I, §2. Each such valuation ν obtained possesses a ramification index e_ν and a local parameter X_ν; for f in $\mathcal{K}$ ν(f) denotes the order of vanishing of the Laurent series of f in powers of X_ν, and $X_\nu = X^{-1/e_\nu}$ if $\nu(X) < 0$, or $X_\nu = (X-a)^{1/e_\nu}$ if $\nu(X-a) > 0$ for some a in $\mathcal{k}$. In the next chapter we shall examine an explicit construction for these Puiseux series, which will be used to establish a bound on the heights of the coefficients and on the genus over k of the field extension they generate.

The ramification indices serve to define the *genus* g of K/k as in (7); we have namely

$$g = 1 - d + \tfrac{1}{2}\sum_{v}(e_v - 1). \tag{21}$$

The genus possesses a spectacular property, termed the Riemann-Roch theorem, which we shall discuss now in preparation for the results on the particular cases $g = 0$ and $g = 1$. We recall from Chapter III, §3 that a divisor $\mathfrak{a}$ on K is defined by a set of components a_v, one integer for each valuation v on K defined above, and with only finitely many a_v non-zero. For each such $\mathfrak{a}$ $d(\mathfrak{a})$ denotes the sum $\sum_v a_v$, and $\ell(\mathfrak{a})$ denotes the dimension over k of the vector space $V(\mathfrak{a})$ consisting of those elements f in K such that $f = 0$ or $(f) + \mathfrak{a} \geq 0$; (f) denotes the principal divisor associated with f, whose component at v is $v(f)$. The Riemann-Roch theorem gives a formula for this dimension, namely

$$\ell(\mathfrak{a}) = d(\mathfrak{a}) + \ell(\mathfrak{w} - \mathfrak{a}) - g + 1, \tag{22}$$

where $\mathfrak{w}$ denotes the divisor on K with components $v(\frac{dX}{dv})$; by (5) has degree $2g-2$. Riemann's theorem, mentioned in §3 of Chapter III, is naturally an immediate consequence of (21), for $\ell(\mathfrak{w} - \mathfrak{a}) \geq 0$, and so $\ell(\mathfrak{a}) \geq 1$ if $d(\mathfrak{a}) \geq g$. We recall that since any principal divisor has degree zero it follows that $\ell(\mathfrak{a}) = 0$ if $d(\mathfrak{a}) < 0$. In §2 we shall see that an immediate consequence of the Riemann-Roch theorem is Luroth's theorem, that if such a field K has genus 0, then it takes the form $k(T)$ for some suitable choice of T in K; the construction of such a uniformising parameter with particular properties is an important part of the algorithm. In §3 we shall show that any field K of genus 1 is elliptic, that is, it takes the form $k(S,T)$, where S and T are connected by a non-degenerate elliptic equation; again, the construction of such a parametrization of K forms an important step.

We recall from Lemma 1 that each element of K may be determined effectively from its principal divisor. By the same method we can construct for each divisor $\mathfrak{a}$ on K a basis of $V(\mathfrak{a})$. This needs no further comment now, but the construction will be examined in greater detail in Chapter V when we shall require an explicit estimate for the heights of the coefficients of such a basis, as did Coates [10] for the classical case of algebraic numbers.

2 *GENUS ZERO*

This section is devoted to the construction of an algorithm which enables all the integral solutions of the general equation of genus zero to be determined effectively. Throughout this section we shall denote by $F(X,Y)$ a polynomial with coefficients in K irreducible over the algebraic closure k of K. We shall assume that the field extension $\mathcal{K}$ of $k(X)$ associated with F has genus zero, and also that it possesses at least three valuations v such that $v(X) < 0$ or $v(Y) < 0$. The following theorems will be established in this section.

Theorem 7

All the solutions X, Y *in* $\mathcal{O}$ *of the equation* $F(X,Y) = 0$ *may be determined effectively.*

Theorem 8

The equation $F(X,Y) = 0$ *possesses an infinite number of solutions in* $\mathcal{O}$ *if and only if there exists a substitution*

$$X = \frac{A(T)}{C(T)}, \quad Y = \frac{B(T)}{C(T)},$$

such that $F(X,Y)$ *vanishes identically in* T; *here* A, B *and* C *are polynomials with no common factor,* A *and* B *have coefficients in* $\mathcal{O}$ *and* C *has coefficients in* k. *Furthermore, if such a substitution exists then there will be the infinite family derived from choosing* T *in* k *with* $C(T)$ *non-zero, but otherwise there will be only finitely many solutions in* $\mathcal{O}$ *of* $F(X,Y) = 0$.

In Chapter V, §2, the following theorem will be proved.

Theorem 9

All the solutions X, Y *in* $\mathcal{O}$ *of* $F(X,Y) = 0$ *satisfy*

$$H(X,Y) \le 40(\Delta+1)^{\Delta^2+11} H + 2\Delta(2g + r - 1);$$

here g *and* r *respectively denote the genus of, and the number of infinite valuations on,* $\mathcal{K}$; H *and* Δ *respectively denote the height and the total degree of* F.

As discussed in §1 we shall show, using the Riemann-Roch theorem, that K is of the form $k(T)$ for some T in K, so X and Y may be expressed as rational functions of T with coefficients in k, and T may be expressed as a rational function of X and Y, again with coefficients in k. Having achieved this construction, we shall employ the fundamental inequality (11) to determine a restriction on the range of possible values for T in k. A detailed examination of these various possibilities will complete the proofs of Theorems 7 and 8. In Chapter V we will calculate a bound on the height of T when X and Y lie in $\mathcal{O}$, and this will lead directly to the bound in Theorem 9 on the heights of solutions in $\mathcal{O}$ of $F(X,Y) = 0$.

Let us begin our analysis by making the observation that the substitution $X = Z + \lambda Y$ transforms F into a polynomial in Z and Y with leading term $F_\Delta(1,\lambda)Y^\Delta$, where F_Δ denotes the homogeneous part of F of degree Δ. Choosing λ in k with $F_\Delta(1,\lambda)$ non-zero, it follows on dividing by $F_\Delta(1,\lambda)$ that F becomes monic of degree Δ with respect to Y. Furthermore, λ may also be chosen so that there are at least three valuations v on K with $v(Z) < 0$, and hence $\Delta \geq 3$. Moreover, for any Y in $\mathcal{O}$, X lies in $\mathcal{O}$ if and only if Z does. We may therefore replace X by Z and so we shall assume henceforth that F is monic of degree Δ with respect to Y and that there exist at least three valuations v on K with $v(X) < 0$. Let us denote one of these valuations by u, and then denote by $\mathfrak{u}$ the divisor of degree 1 on K with a single non-zero component, at u. We may now apply the Riemann-Roch theorem (22) to conclude that

$$\ell(\mathfrak{u}) = 2$$

since $d(\mathfrak{w} - \mathfrak{u}) = -3 < 0$. We recall from §1 that a k-basis of $V(\mathfrak{u})$ may be determined effectively; it will consist of 1 and T say, where T lies in $K \setminus k$ and is expressed in the form $\sum_{i=1}^{\Delta} A_i Y^{i-1}$, where $A_1, \ldots, A_\Delta$ are computable rational functions, in $k(X)$. Now T has a single pole of order 1, and so $H(T) = 1$, where

$$H(f) = -\sum_{v} \min(0, v(f))$$

for any f in K. However, as in Chapter I we have $H(f) = [K:k(f)]$ for f in $K \setminus k$, so $K = k(T)$ as required. It follows that X and Y may be expressed as rational functions of T with coefficients in k. In fact $X = P(T)/\alpha Q(T)$, where $P(T) = \prod (T - T_\nu)^{e_\nu}$, the product being taken over all the valuations ν on K with $\nu(X) > 0$; $e_\nu = \nu(X)$, so P has degree Δ, and T_ν denotes the unique element of k with $\nu(T - T_\nu) > 0$, namely the constant coefficient of the Puiseux series of T at ν, which may be calculated from the expression for T in terms of X and Y above. Similarly $Q(T) = \prod (T - T_\nu)^{e_\nu}$, where the product runs over all the valuations $\nu \neq u$ such that $\nu(X) < 0$; $e_\nu = -\nu(X)$, so Q has degree $\Delta - e_u$, and T_ν lies in k with $\nu(T - T_\nu) > 0$ as above; since there are at least two such valuations ν we conclude that Q has at least two distinct zeros. Hence the polynomials P and Q may be effectively determined, whilst α may be computed by calculating a value of T corresponding to $X = 1$, say. We now denote by L the finite extension field of K obtained by adjoining the coefficients of $A_1, \ldots, A_\Delta$ and P, together with α and the zeros of Q. Similarly the rational function A such that $Y = A(T)$ may be determined effectively, and we observe that since Y/X is integral over $k[1/X]$, the denominator of A divides Q in $k[T]$.

So far X, Y and T have been elements of K, that is, variables over k. We now wish to determine all the possible values of X and Y in $\mathcal{O}$ such that $F(X,Y) = 0$. This will be achieved by investigating the corresponding values of T in k and successively restricting its permissible range. Let us first consider the case when X lies at a pole of one of the rational functions $A_1, \ldots, A_\Delta$. Here X has only finitely many possibilities, and for each of these we may determine the corresponding values of Y such that $F(X,Y) = 0$. We may then select those solutions X, Y which actually lie in $\mathcal{O}$. Henceforth we shall assume that X, Y is a solution in $\mathcal{O}$ of $F(X,Y) = 0$ such that none of $A_1, \ldots, A_\Delta$ has a pole at X. Now the corresponding value of T is equal to $\sum_{i=1}^{\Delta} A_i(X) Y^{i-1}$, and so it lies in the field L; we wish to determine all its possible values. Let β and γ denote two distinct zeros of Q; we thus have

$$X\alpha(T - \beta)(T - \gamma)R(T) = P(T) , \tag{23}$$

where R(T) is the quotient of Q(T) by $(T-\beta)(T-\gamma)$. This equation, together with

$$(T - \beta) + (\gamma - T) + (\beta - \gamma) = 0,$$

to which we shall apply the fundamental inequality, will suffice to restrict severely the range of permissible values of T in L. In Chapter VI these two equations will be shown to lead directly to a bound on the height of T in L (Lemma 9). Let us here denote by $\mathcal{V}$ the set of valuations v on L at which one or more of the following occur:

$$v|\infty,\ v(\alpha)<0,\ v(Q)<0,\ v(P)<0,\ v(\beta-\gamma)>0,\ v(P(\beta))>0,\ v(P(\gamma))>0.$$

Now $\mathcal{V}$ is a finite computable set, and we claim that whenever v lies outside $\mathcal{V}$, and T lies in L, and X lies in $\mathcal{O}$, then

$$v(T-\beta) = v(T-\gamma) = v(\beta-\gamma) = 0.$$

To prove this assertion, we first observe that since $\deg P = \Delta > \deg Q$, (23) represents a monic equation for T with all its poles in $\mathcal{V}$, so $v(T) \geq 0$ for v outside $\mathcal{V}$. Similarly $v(\beta) \geq 0$ and $v(\gamma) \geq 0$ for v outside $\mathcal{V}$. Moreover if $v(T-\beta) > 0$ and $v(P(T)) > 0$, then $v(P(\beta)) > 0$, so v lies in $\mathcal{V}$; similarly $v(T-\gamma) = 0$ for v outside $\mathcal{V}$, and so our assertion is proved. We may now apply the corollary in Chapter I to Lemma 2, and conclude that either $(T-\beta)/(\beta-\gamma)$ lies in the ground field k, or it belongs to some finite effectively determinable subset of L. In the latter case each of the finitely many possibilities for T may be substituted into the expressions $X = P(T)/\alpha Q(T)$, $Y = A(T)$ to achieve a finite set of solutions in L of $F(X,Y) = 0$; those which actually lie in $\mathcal{O}$ may then be selected. If $Q(T)$ vanishes at one of these values for T, then X has a pole, so such values may be dismissed a priori.

Henceforth we shall assume that $S = (T-\beta)/(\beta-\gamma)$ is an element of k, and we wish to determine its range of possible values corresponding to X and Y in $\mathcal{O}$. If we first replace T by S, then we may in fact assume that T lies in k, $\beta = 0$ and $\gamma = 1$. We shall prove that unless Q has coefficients in k and P/α has coefficients in $\mathcal{O}$ then T has only finitely many possibilities in k which may be determined effectively. Let us first suppose that Q has a zero δ outside the ground field k. If we denote by $\mathcal{W}$ the set of valuations v on L for which

$$v|\infty,\ v(\alpha) < 0,\ v(Q) < 0,\ v(P) < 0,\ v(P(\delta)) > 0,$$

then as above $\mathcal{W}$ is a finite computable set and $v(T-\delta) = 0$ for v outside

$\mathcal{W}$. However, if T lies in k but δ does not, then $0 < H_L(\delta) = H_L(T-\delta) = \sum_v \max(0, v(T-\delta))$, where the sum runs over all valuations v on L. Hence $v(T-\delta) > 0$ for some valuation v on L, so v lies in $\mathcal{W}$, and thus T is equal to the constant coefficient $\delta(v)$ of the Puiseux expansion of δ at one of the finitely many valuations v in $\mathcal{W}$ with $v(\delta) \geq 0$. We conclude that if Q(T) has a zero outside k then there are only finitely many possibilities for T in k, each of which may be determined effectively, as required.

We shall assume henceforth that Q has coefficients in k; since k is algebraically closed Q also has its roots in k. Let us denote by N(T) the polynomial $P(T)/\alpha$, so $XQ(T) = N(T)$. We shall prove that if T has infinitely many possibilities in k then N has coefficients in $\mathcal{O}$, and that otherwise the finitely many possibilities for T may be determined effectively. Let us first denote by $x_1(=1), x_2, \ldots, x_m$ a K-basis of L, so that each f in L may be written uniquely in the form $\sum_{i=1}^{m} f_i x_i$ with $f_1, \ldots, f_m$ in K. We may thus write $N(T) = \sum_{i=1}^{m} N_i(T) x_i$, where $N_1, \ldots, N_m$ are polynomials with coefficients in K. Since X lies in K, and T lies in k as do the coefficients of Q, we have $N_2(T) = \ldots = N_m(T) = 0$. Hence unless $N_2, \ldots, N_m$ all vanish identically and so N has coefficients in K, then T has only finitely many possibilities in k which may be determined effectively, as required.

Let us now assume that Q has coefficients in k and N has coefficients in K, but that not all the coefficients lie in $\mathcal{O}$; we wish to show that T has only finitely many possibilities in k which may be computed. In this case there exists a finite valuation v on K with $v(N) < 0$, so if σ_v denotes the embedding of K in the field of Laurent series in powers of z_v, we have

$$\sigma_v(N(T)) = \sum_{h=v(N)}^{\infty} a_h(T) z_v^h ,$$

where each a_h is a polynomial in T with coefficients in k, and $a_{v(N)}(T)$ is not identically zero. Since X lies in $\mathcal{O}$ and T lies in k with $Q(T) \neq 0$, we have $v(Q(T)) = 1$ since Q(T) is a non-zero element of k; hence $a_{v(N)}(T) = 0$, so T has only finitely many possibilities in K.

We conclude that unless X may be expressed as $N(T)/Q(T)$ with N in $\mathcal{O}[T]$ and Q in $k[T]$, then T has only finitely many possibilities in k such that X lies in $\mathcal{O}$, and these possibilities are effectively determinable. A similar conclusion obtains for the expression for Y as a rational function of T. If X and Y can be so expressed, then T may assume any value in k such that $Q(T) \neq 0$, and the corresponding values of X and Y evidently form a solution in $\mathcal{O}$ of $F(X,Y) = 0$. In any case, T has only finitely many possibilities in $k \setminus k$, and these may be determined effectively; hence the proofs of Theorems 7 and 8 are complete.

We conclude this section with three remarks. First, when $F(X,Y) = 0$ has infinitely many solutions in $\mathcal{O}$ it follows immediately that T may be expressed in the form $\sum_{i=1}^{\Delta} A_i Y^{i-1}$, where each A_i is a rational function in $K(X)$, not just in $k(X)$. For we may select a common denominator $E(X)$ for $A_1,\ldots,A_\Delta$ and so write $A_i(X) = A_i(X)/E(X)$ for each i, where E, $A_1,\ldots,A_\Delta$ have coefficients in L. As above we may write each of these polynomials as $E = \sum_{j=1}^{m} E_j x_j$, $A_i = \sum_{j=1}^{m} A_{ij} x_j$, where each E_j and A_{ij} has coefficients in K. Since E is non-zero, some E_j is non-zero, and if T lies in k we obtain

$$TE_j(X) = \sum_{i=1}^{\Delta} A_{ij}(X) Y^{j-1},$$

which, on substituting $X = N(T)/Q(T)$, $Y = A(T)$, leads to an equation $G(T) = 0$ for T. However, T may assume any value in k such that $Q(T)$ is non-zero, so G must vanish identically; hence T is expressed in the form $\sum_{i=1}^{\Delta} B_i(X) Y^{j-1}$, where the rational functions $B_i = A_{ij}/E_j$ have coefficients in K as required.

Secondly we observe that as in Siegel's original paper [40] the requirement that Y is integral may be removed. Let us suppose that $F(X,Y)$ is a polynomial with coefficients in K which is irreducible over the algebraic closure k of K, such that there are at least three valuations v on the associated field extension K of $k(X)$ with $v(X) < 0$, and such that K/k has genus zero. We conclude as above that all the solutions X in $\mathcal{O}$, Y in K, of $F(X,Y) = 0$ may be determined effectively. In general there will be only finitely many such solutions. The only case to the contrary occurs when there is a non-trivial substitution

$$X = \frac{A(T)}{B(T)}, \quad Y = \frac{C(T)}{D(T)} \qquad (A \in \mathcal{O}[T], B \in k[T], C, D \in K[T])$$

such that F vanishes identically in T. Then there will be the infinite family of solutions obtained from choosing T in k with $B(T) \neq 0 \neq D(T)$, but otherwise only finitely many solutions X in $\mathcal{O}$, Y in K.

Finally, if K' is any finite extension field of k(z) then all the solutions X in $\mathcal{O}$, Y in K' of F(X,Y) = 0 may be determined effectively, and infinitely many such solutions exist only when C and D above may be chosen to have coefficients in K'.

3 GENUS ONE

This section will be devoted entirely to the construction of an algorithm whereby may be determined all the solutions of any equation of genus one which are integral in an arbitrary finite extension K of k(z). Our approach is analogous to that of Baker and Coates [8] in the classical case of number fields, and it is to transform the equation into an elliptic equation by means of a birational transformation; this elliptic equation may be solved as in Chapter II. In the next chapter an examination of the parameters involved at each stage in the construction of such an elliptic equation will serve to establish a bound on the heights of the solutions in $\mathcal{O}$ of the equation of genus one. This bound (Theorem 12) is a linear function of the height of the original equation, which provides a contrast with the exponential bounds obtained in the classical case. Let F(X,Y) denote a polynomial in X and Y with coefficients in K, irreducible over the algebraic closure $\mathfrak{k}$ of K and such that the associated field $\mathcal{K} = \mathfrak{k}(X,Y)$ has genus 1 over $\mathfrak{k}$. Here we shall establish the following theorems.

Theorem 10

All the solutions X, Y *in* $\mathcal{O}$ *of the equation* F(X,Y) = 0 *may be determined effectively.*

Theorem 11

In general F(X,Y) = 0 *has only finitely many solutions in* $\mathcal{O}$. *The only case to the contrary occurs when there is a non-trivial substitution*

$$X = \frac{A(T) + SB(T)}{C(T)}, \quad Y = \frac{D(T) + SE(T)}{G(T)},$$

where A, B, D *and* E *are polynomials with coefficients in* $\mathcal{O}$, C *and* G *are polynomials with coefficients in* k, *such that* F(X,Y) *vanishes whenever*

$$S^2 = J(T) ,$$

where J *is a monic cubic polynomial with distinct roots in* k. *Moreover, in this case there will be the infinite family obtained by choosing* T *in* k *with* $C(T) \neq 0 \neq G(T)$, *but otherwise only finitely many solutions* X, Y *in* $\mathcal{O}$ *of* $F(X,Y) = 0$.

In Chapter V we shall prove the following theorem.

Theorem 12

All the solutions X, Y *in* $\mathcal{O}$ *of* $F(X,Y) = 0$ *satisfy*

$$H(X,Y) \le 1500(\Delta+1)^{\Delta^2+11} H + 8\Delta(2g + r - 1),$$

where g *and* r *respectively denote the genus of* K/k *and the number of infinite valuations on* K; H *and* Δ *respectively denote the height and total degree of* F.

The establishment of Theorems 10 and 11 commences with the construction of a birational substitution whereby $F(X,Y) = 0$ is transformed into an elliptic equation. Let us denote by u a fixed valuation on K at which X has a pole, and then denote by $\mathfrak{u}$ the divisor on K of degree 1 with a single non-zero component, at u. By the Riemann-Roch theorem (22) we have $\ell(n\mathfrak{u}) = n$ for $n \ge 1$, since $d(\mathfrak{w}) = 0$. Hence we obtain $\ell(n\mathfrak{u}) - \ell(n\mathfrak{u} - \mathfrak{u}) = 1$ for $n \ge 2$, and thus there exist U and V in K with poles of orders 2 and 3 respectively at u, and no other poles. Furthermore, as we observed in §1, the method of Lemma 1 enables us to construct such elements in the form $\sum_{i=1}^{d} A_i Y^{i-1}$, where $A_1, \ldots, A_d$ are computable rational functions of X with coefficients in k; d denotes the degree of K over $k(X)$, that is, the degree of F in Y. As in the case of genus 0 we may replace X by $X + \lambda Y$ for suitable choice of λ in k and so assume that F is monic of degree Δ with respect to Y; this assumption suffices for the proofs of Theorems 10 and 11. In fact it will be of greater significance when we show how the Puiseux expansions may be determined effectively (Lemma 5); these, and the actual construction of the functions U and V, are somewhat simplified by this assumption on F. Now the 7 elements 1, U, V, U^2, UV, U^3, V^2 have poles at u of orders 0, 2, 3, 4, 5, 6, 6 respectively, and no other poles. However $\ell(6\mathfrak{u}) = 6$, so there exist $\delta_1, \ldots, \delta_7$ in k such that not all δ_i are zero, but

$$W = \delta_1 + \delta_2 U + \delta_3 V + \delta_4 U^2 + \delta_5 UV + \delta_6 U^3 + \delta_7 V^2$$

vanishes identically. Furthermore, such elements may be determined effectively, for $\nu(W) \geq 0$ for $\nu \neq u$, and $u(W) \geq -6$ regardless of $\delta_1, \ldots, \delta_7$ so it suffices to solve the equations

$$\delta_1 + \sum_{i=2}^{7} \delta_i h_{i0} = 0 , \quad \sum_{i=2}^{7} \delta_i h_{ij} = 0 , \quad j=-1, \ldots, -6,$$

where U, V, U^2, UV, U^3, V^2 have Puiseux expansions $\sum_{j=-6}^{\infty} h_{ij} X_u^j$ $i=2,\ldots,7$ respectively. Since $\ell(5u) = 5$, no linear relation exists between 1, U, V, U^2 and UV, so we have $\delta_6 \delta_7 \neq 0$ as U^3 and V^2 have poles of order 6 at u; thus we may assume that $\delta_7 = 1$. Upon replacing V by $V + \frac{1}{2}\delta_5 U + \frac{1}{2}\delta_3$ we obtain the equation

$$V^2 = h(U - h_1)(U - h_2)(U - h_3) , \tag{24}$$

where h, h_1, h_2, h_3 are elements of k and h is non-zero. Furthermore, if $h_1 = h_2$ say then $V/(U-h_1)$ would have a single pole, of order 1, at u, in contradiction to the Riemann-Roch theorem: hence h_1, h_2, h_3 are distinct. Now $H(U) = -\sum_{\nu} \min(0, \nu(U)) = 2$, so $[K:k(U)] = 2$; similarly $[K:k(V)] = 3$, and so $K = k(U, V)$. We may thus express X and Y in the form $(\Xi(U)+V\Psi(U))/\Omega(U)$, where Ξ, Ψ and Ω are polynomials in U with coefficients in k. It is necessary actually to determine such polynomials Ξ, Ψ and Ω, and so we now proceed to show how this may be accomplished. If $\nu \neq u$ is a valuation on K then we denote by U_ν the constant coefficient of the Puiseux expansion of U at ν, so $\nu(U - U_\nu) > 0$; in fact $\nu(U - U_\nu) = 1$ if $U_\nu \neq h_1, h_2, h_3$, otherwise $\nu(U - U_\nu) = 2$. Now if $\Omega(U) = \prod (U - U_\nu)^{e_\nu}$, where the product is taken over all the valuations $\nu \neq u$ on K such that $\nu(X) < 0$, then since $\nu(X) = -e_\nu$ for such a valuation, we deduce that $X\Omega(U)$ has a single pole, of order $2\Delta - e_u$, at u. However, 1 and V form a $k[U]$ basis of the ring of elements Z in K with $\nu(Z) \geq 0$ whenever $\nu \neq u$, so it follows that $X\Omega(U)$ may be expressed in the form $\Xi(U) + V\Psi(U)$ for some polynomials $\Xi(U)$, $\Psi(U)$ with coefficients in k. If we compare the orders of the poles at u then we have

$$\max\{2 \deg \Xi, 3+2 \deg \Psi\} = -u(\Xi(U)+V\Psi(U)) = -u(X\Omega(U)) = 2\Delta - e_u,$$

and so the degrees of Ξ and Ψ are bounded. As in the determination of $\delta_1,\ldots,\delta_7$ above, we have $v(\Xi(U)+V\Psi(U)-X\Omega(U)) \geq 0$ for all valuations $v \neq u$, so it suffices to solve the $2\Delta+1-e_u$ linear equations in the coefficients of Ξ and Ψ which express that $\Xi(U)+V\Psi(U)-X\Omega(U)$ has a zero at u; it must then vanish identically as required. Hence we may express X effectively in the form

$$X = \frac{\Xi(U)+V\Psi(U)}{\Omega(U)}$$

for some polynomials Ξ, Ψ, Ω in $k[U]$; clearly it may also be assumed that these polynomials have no common factor. Similarly we may effectively express Y in this form. Eliminating V we obtain

$$\Xi^2(U) - V^2\Psi^2(U) - 2X\Xi(U)\Omega(U) + X^2\Omega^2(U) = 0 ,$$

where V^2 is the cubic polynomial in U given by (24) above. The polynomial on the right hand side of this expression is of degree $e_u+2\deg\Omega$ in U, and so, since $e_u \geq 1$, the leading term in U does not involve X. After multiplying throughout by some factor in k such that all the coefficients become integral over $k[z]$, let this leading coefficient be denoted by c. Thus whenever X assumes a value in k integral over $k[z]$, so does cU. Let us now denote by a some element of k such that a/c, ah_1, ah_2, ah_3 are all integral over $k[z]$, and by b some element such that $b^2 = a^3/h$. The substitution $T = aU$, $S = bV$ transforms (24) into

$$S^2 = (T-\alpha_1)(T-\alpha_2)(T-\alpha_3) , \tag{25}$$

where α_1, α_2 and α_3 are distinct elements of k integral over $k[z]$. Furthermore, the coefficients in the birational transformation $(X,Y) \leftarrow - \rightarrow (T,S)$ are effectively determinable, that is, we may compute rational functions $A_1,\ldots,A_\Delta,B_1,\ldots,B_\Delta$ in $k(X)$ and polynomials A, B, C, D, E, G in $k[T]$ such that

$$T = \sum_{i=1}^{\Delta} A_i(X)Y^{i-1}, \quad S = \sum_{i=1}^{\Delta} B_i(X)Y^{i-1} ,$$
$$X = \frac{A(T)+SB(T)}{C(T)}, \quad Y = \frac{D(T)+SE(T)}{G(T)} ; \tag{26}$$

furthermore, it may be assumed that C and G are monic and each of A, B, C and D, E, G have no common factor. Let us denote by L the finite extension of K generated by all the coefficients in (26) together with $\alpha_1, \alpha_2, \alpha_3$.

Until now X, Y, T and S have been elements of K, that is, variables over k. We shall now regard X and Y as elements of $\mathcal{O}$ satisfying $F(X,Y)=0$. By determining all possible values for the corresponding elements T and S in k we shall provide an algorithm which furnishes the complete list of solutions X, Y in $\mathcal{O}$ as required. Now if X lies at a pole of a rational function $A_1,\ldots,A_\Delta,B_1,\ldots,B_\Delta$, then X has only finitely many possibilities. For each of these we may determine the corresponding values of Y such that $F(X,Y)=0$, and then select from these pairs the solutions which actually lie in $\mathcal{O}$. Henceforth we shall assume that the rational functions $A_1,\ldots,A_\Delta,B_1,\ldots,B_\Delta$ are regular at X, and so, if X and Y lie in K, then T and S lie in L. Furthermore, as noted above, if X is integral over $k[z]$ then so are T and S, and thus if X and Y both lie in $\mathcal{O}$ then T and S lie in $\mathcal{O}_L$, the ring of elements of L integral over $k[z]$. We may now apply the algorithm established in Chapter III and thereby determine all the solutions T, S in $\mathcal{O}_L$ of the elliptic equation (25) above. If this elliptic equation has only finitely many solutions in $\mathcal{O}_L$, then, having determined these, we may compute from (26) the corresponding values of X and Y, and select just those solutions in $\mathcal{O}$. For T lying at a zero δ of C or G the corresponding values of X and Y are found from developing the Puiseux series of S at $T=\delta$. Henceforth we shall assume that (25) has an infinite number of solutions T, S in $\mathcal{O}_L$. By performing the integral linear substitution specified in Theorem 5 we may in fact assume that $\alpha_1, \alpha_2, \alpha_3$ are elements of the ground field k. The finite number of solutions of (25) not in the infinite family with T, S in k may each be dealt with separately as above. It remains only to be determined which values of T, S in k lead to a solution X, Y in $\mathcal{O}$ of $F(X,Y)=0$. Clearly if A, B, D, E have coefficients in $\mathcal{O}$, and C, G have coefficients in k then T may assume any value in k such that $C(T)\neq 0\neq G(T)$; X and Y are then determined from (25) and (26) and they will be elements of $\mathcal{O}$ for both choices of S in k. We shall show that if these conditions are not satisfied then T has only finitely many possibilities in k which may be determined effectively. This will complete the proofs of Theorems 10 and 11, as each of these finitely many possibilities may be dealt with separately, as above.

Let us first suppose that not all the coefficients of the polynomial C lie in k, so that C has a non-constant zero α; let $L(\alpha)$ denote the field obtained by adjoining α to L. We shall show that there are now only finitely many computable possibilities for T in k. There are in fact two cases to be considered, according as α is not or is a zero of the polynomial $N(T) = A^2(T)-S^2B^2(T)$, where S^2 is the polynomial in T given by (25). If $\beta = N(\alpha)$ is non-zero, then we may consider the equation

$$N(T) = X\, C(T)(A(T) - SB(T)).$$

Let $\mathcal{V}$ denote the set of valuations v on $L(\alpha)$ at which $v|\infty$, $v(A,B,C) < 0$ or $v(\beta) > 0$; such a set $\mathcal{V}$ is both finite and effectively determinable. We deduce that if T lies in k and $v(T-\alpha) \neq 0$ then v lies in $\mathcal{V}$, as otherwise $v(T-\alpha) > 0$ and so $v(\beta) > 0$. However, since α is non-constant we deduce that $v(T-\alpha) > 0$ for some v, so v lies in $\mathcal{V}$ and T must be equal to the constant coefficient of the Puiseux expansion of α at one of the finitely many valuations v in $\mathcal{V}$ for which $v(\alpha) \geq 0$. Hence if α is not a zero of N then T has only finitely many effectively determinable possibilities in k as required. Let us now assume that α is a zero of N. Since A, B, C have no common factor there exist non-zero elements γ and δ in $L(\alpha)$ such that $B(\alpha) = \delta$ and $A(\alpha) = -\gamma\delta$; $N(\alpha) = 0$ and so $\gamma^2 = (\alpha-\alpha_1)(\alpha-\alpha_2)(\alpha-\alpha_3)$. Let $\mathcal{W}$ denote the set of valuations v on $L(\alpha)$ at which $v|\infty$, $v(A,B,C) < 0$ or $v(\delta) > 0$; such a set $\mathcal{W}$ is both finite and effectively determinable. Now

$$XC(T) = A(T) + SB(T) ,$$

so if T and S are elements of k and X is an element of $\mathcal{O}$, and v is a valuation outside $\mathcal{W}$ with $v(T-\alpha) > 0$, then $v(S-\gamma) > 0$. However α is non-constant, so β is not in $k(\alpha)$, and thus there is a valuation v on $L(\alpha)$ with $v(T-\alpha) > 0$ and $v(S+\gamma) > 0$, and from the above either $S = 0$ or v lies in $\mathcal{W}$. Hence T is equal to α_1, α_2, α_3 or the constant coefficient of the Puiseux expansion of α at one of the finitely many valuations v in $\mathcal{W}$ with $v(\alpha) \geq 0$. We conclude that unless C has all its roots in the ground field k, then T has only finitely many possibilities in k which may be determined effectively. A similar result may be obtained for the polynomial G; we shall assume henceforth that C and G have coefficients, and hence roots, in k.

We shall now show that unless the coefficients of A and B lie in $\mathcal{O}_L$, then there are only a finite number of possibilities for T in k such that X lies in $\mathcal{O}$. For let us suppose that there is a finite valuation v on L such that $v(A,B) < 0$. Expressing the coefficients of A and B as Puiseux series in the local parameter z_v we obtain

$$\sigma_v(A(T)) = \sum_{h=v(A)}^{\infty} a_h(T) z_v^h, \quad \sigma_v(B(T)) = \sum_{h=v(B)}^{\infty} b_h(T) z_v^h,$$

where the polynomials a_h, b_h have coefficients in k. Since C has coefficients in k and T lies in k we have $C(T) = 0$ or $v(C(T)) = 1$. In the latter case we obtain $a_h(T) + Sb_h(T) = 0$ for $h < 0$, since $v(X) \geq 0$ and S also lies in k. Substituting for S we obtain

$$C(T) = 0 \quad \text{or} \quad a_h^2(T) - (T-\alpha_1)(T-\alpha_2)(T-\alpha_3)b_h^2(T) = 0 \qquad (h<0).$$

The latter equation is non-trivial for $h = v(A,B)$, and so T has only finitely many possibilities in k which may be determined effectively, as required. A similar result applies to the polynomials D and E: unless they have coefficients in $\mathcal{O}_L$ there are only a finite number of possibilities for T in k such that Y lies in $\mathcal{O}$.

Finally we show that unless A and B have coefficients in K then T has only finitely many possibilities in k. As in §2 we denote by $x_1(=1), x_2, \ldots, x_m$ a K-basis of L, so that each f in L may be written uniquely in the form $\sum_{i=1}^{m} f_i x_i$ with $f_1, \ldots, f_m$ in K. Hence we may write $A(T) = \sum_{i=1}^{m} A_i(T)x_i$ and $B(T) = \sum_{i=1}^{m} B_i(T)x_i$, with $A_1, \ldots, A_m, B_1, \ldots, B_m$ polynomials with coefficients in K. Since $XC(T) = A(T)+SB(T)$, and $XC(T)$ lies in K if T lies in k, we deduce that $A_i(T)+SB_i(T) = 0$ for $2 \leq i \leq m$. As above we obtain a non-trivial equation for T unless $A_2, \ldots, A_m, B_2, \ldots, B_m$ all vanish identically, that is, unless A and B have coefficients in K as required. Similarly we conclude that unless D and E have coefficients in K then T has only finitely many possibilities in k which may be determined effectively, as required.

As in §2 we conclude with three remarks. First, when $F(X,Y) = 0$ has infinitely many solutions in $\mathcal{O}$ it follows that T and S may be expressed in the form $\sum_{i=1}^{\Delta} A_i Y^{i-1}$ and $\sum_{i=1}^{\Delta} B_i Y^{i-1}$ respectively, where each

A_i, B_i is a rational function of X with coefficients in K, not just in k. The proof of this result is just as in §2. Secondly, as in Siegel's paper *[40]* the requirement that both X and Y are integral may be removed. If F is as in Theorems 10 and 11, then we may determine effectively all the solutions X in $\mathcal{O}$, Y in K of $F(X,Y) = 0$. We also conclude that there are in general only finitely many solutions, the only case to the contrary occurring when there is a non-trivial substitution

$$X = \frac{A(T) + SB(T)}{C(T)}, \quad Y = \frac{D(T) + SE(T)}{G(T)} \; (A,B \; \mathcal{O}[T], C \; k[T], D,E,G \; K[T])$$

such that $F(X,Y)$ vanishes whenever

$$S^2 = J(T),$$

where J is a monic cubic polynomial with distinct roots in k. In this case there will be the infinite family obtained by choosing T in k with $C(T) \neq 0 \neq G(T)$, but otherwise only finitely many solutions X in $\mathcal{O}$, Y in K of $F(X,Y) = 0$.

Finally, if K' is any finite extension of $k(z)$ then all the solutions X in $\mathcal{O}$, Y in K' of $F(X,Y) = 0$ may be determined effectively, infinitely many such solutions existing only when D, E and G above may be chosen to have coefficients in K'.

CHAPTER V BOUNDS FOR EQUATIONS OF SMALL GENUS

1 *PRELIMINARIES*

In this chapter we shall expand the results obtained in Chapter IV on the complete resolution of equations of genera 0 and 1 by determining explicit bounds on the heights of all their integral solutions, as expressed in Theorems 9 and 12. It is to be remarked that these bounds are linearly dependent on the height of the equation concerned, in contrast with the classical case when the bounds established by Baker and Coates *[8]* are of multiply exponential growth. Our method of proof consists of a detailed analysis of the construction of the algorithms derived in Chapter IV, coupled with an estimation of the various parameters involved at each stage thereof. Central to the constructions are Puiseux's theorem (see Chapter I) and the Puiseux expansions; in this section we shall establish the requisite bounds on the coefficients in any Puiseux expansion. First, however, we shall require a bound on the genus of any finite extension of $k(z)$. Throughout this chapter we shall denote by L a sufficiently large finite extension of K, and, unless otherwise stated, for f in L $H(f)$ will denote the sum $-\sum \min(0,v(f))$ taken over all the valuations v on L. If K' is any field lying between K and L then we denote by $G_{K'}$ the integer $[L:K'](g_{K'}-1)$, where $g_{K'}$ is the genus of K'/k and $[L:K']$ is the degree of L over K'; we also recall that the height in K' of any element f is given by $H_{K'}(f) = H(f)/[L:K']$.

Lemma 4

Let $A(T) = T^n + \alpha_1 T^{n-1} + \ldots + \alpha_n$ *denote a polynomial which is irreducible over* K. *If* K' *is the field obtained by adjoining a zero* t *of* A *to* K, *then we have*

$$H(t) = H/n \quad \textit{and} \quad 0 \le G_{K'} - G_K \le \frac{3}{2}(1-1/n)H,$$

where H *denotes the height of* A.

Proof. By assumption L is sufficiently large that A factorises completely in L; let the zeros of A in L be $t_1(=t), t_2, \ldots, t_n$. From Chapter I we obtain $\sum_{i=1}^{n} H(t_i) = H$, and, since $t_1, \ldots, t_n$ are conjugate over K, they have equal height. The first assertion is thus established, and to prove the second we note that, from the genus formula (7), we obtain

$$G_{K'} - G_K = \tfrac{1}{2}[L:K']\sum_{v}(\varepsilon_v - 1), \tag{27}$$

where the sum extends over all the valuations v on K', and ε_v denotes the relative ramification index of v over K; thus $G_{K'} \geq G_K$. We shall bound separately the contributions to the sum on the right of (27) from the valuations v on K' according as $v(A) < 0$ or $v(A) \geq 0$.

If $v(A) < 0$, then $v|w$ for some valuation w on K with $w(A) < 0$. Since the sum of the positive integers $-w(A)$ over such valuations is just the height $H_K(A)$ of A with respect to K, we conclude that there are at most $H_K(A)$ such valuations w on K. Furthermore, for each w we have $\sum_{v|w} \varepsilon_v = [K':K] = n$, and so the valuations v with $v(A) < 0$ contribute at most

$$\tfrac{1}{2}[L:K']H_K(A)(n-1) = \tfrac{1}{2}[L:K]H_K(A)(1-1/n) = \tfrac{1}{2}(1-1/n)H$$

to the sum on the right hand side of (27).

It now remains to prove that the contribution from valuations v with $v(A) \geq 0$ is at most $(1-1/n)H$. Let us denote by ν a valuation on L such that $\nu|v$, so whenever f lies in K' we have $\nu(f) = e_\nu v(f)$ for some fixed integer e_ν; we note that $\sum_{\nu|v} e_\nu = [L:K']$ for each v. Now $v(A) \geq 0$, so $\nu(t_i) \geq 0$ for each i, and thus the expansion of t is regular at v, given by $\sigma_v(t) = \sum_{h=0}^{\infty} c_h z_v^h$ say, where each c_h lies in k. As ζ ranges over the ε_v-th roots of unity, the ε_v series $\sum_{h=0}^{\infty} c_h \zeta^h z_v^h$ are all distinct zeros of A, and so must be the expansions of some of $t_1, \ldots, t_n$. However, if t_i, $i>1$, is one such, then $\nu(t-t_i) \geq \nu(z_v)$, since t and t_i have the same constant coefficient. Hence $\sum_{i=2}^{n} \nu(t-t_i) \geq e_\nu(\varepsilon_v - 1)$ since $e_\nu = \nu(z_v)$, and so the contribution to the sum on the right of (27) from valuations v on K' with $v(A) \geq 0$ is at most

$$\tfrac{1}{2}[L:K']\sum_{v}(\varepsilon_v-1) = \tfrac{1}{2}\sum_{v}\sum_{\nu|v} e_\nu(\varepsilon_v-1) \le \tfrac{1}{2}\sum_{\nu}\sum_{i=2}^{n}\nu(t-t_i)$$

$$\le \tfrac{1}{2}\sum_{i=2}^{n} H(t-t_i) \le (1-1/n)H$$

as required, where each sum extends over those valuations v,ν with $\nu|v$, $v(A)\ge 0$. The proof of the lemma is thus complete, and we note that since $\frac{\partial A}{\partial T}(t) = \prod_{i=2}^{n}(t-t_i)$, any valuation v on K' with $\varepsilon_v > 1$ satisfies $v(A) < 0$ or $v(\frac{\partial A}{\partial T}(t)) > 0$.

The condition that A is irreducible may be removed.

Corollary

Let $A(T)$ *denote a non-zero polynomial of degree* n *with coefficients in* K. *If* K' *is the field obtained by adjoining a zero* t *of* A *to* K, *then we have*

$$[K':K] \le n, \quad [K':K]H(t) \le H \quad \textit{and} \quad 0 \le G_{K'}-G_K \le \tfrac{3}{2}(1-1/n)H,$$

where H *is the height of the polynomial* A.

Proof. Let us denote by $B(T)$ the product of factors $T-x$ as x runs over the conjugates of t over the field K; thus B is an irreducible polynomial over K of which t is a zero, and B has degree $[K':K]$. Furthermore, since B is a divisor of A we conclude that B has degree at most n and height at most H. The result now follows on applying Lemma 4 to the polynomial B.

We now wish to estimate both the heights of the coefficients in a Puiseux series and the increase in genus obtained as a result of the adjunction of these coefficients. The results will first be derived for a power series and then extended to a general Puiseux series; the method of proof is substantially that of Coates *[9]*.

Lemma 5

Let $F(X,Y)$ *denote a polynomial with coefficients in* K, *which is separable with respect to* Y. *Suppose that if* $Y_0 = \sum_{i=0}^{\infty}\alpha_i X^i$, *where each* α_i *lies in* k, *then* $F(X,Y_0)$ *represents the zero power series in* X; *let* m *be defined as the order of vanishing in* X *of the series* $\frac{\partial F}{\partial Y}(X,Y_0)$. *Then*

$$H(\alpha_i) \le (\Delta+1)^i H \qquad (0 \le i \le m),$$

$$H(\alpha_j;\ m<j\le i) \le (2i-2m-1)(\Delta+1)^{m+1} H \qquad (i>m),$$

where H *denotes the height of* F, *and* Δ *the degree of* F *with respect to* Y. *Furthermore, we have*

$$G_{K'} - G_K \le \frac{3}{2}(\Delta+1)^m H,$$

where K' *is the field obtained from* K *by adjoining all the coefficients* α_i, $i \ge 0$.

Proof. We first observe that, since F is separable with respect to Y, we have $\frac{\partial F}{\partial Y}(X,Y_0) \ne 0$ and so m is finite. Let us write $K_0 = K$, and denote by K_i, $i \ge 1$, the field obtained by adjoining to K the coefficients $\alpha_0,\dots,\alpha_{i-1}$. We shall show that $K_{m+1} = K'$, so K' is a finite extension of K as required; in fact it is evident that $[K':K] \le \Delta$. To prove the first part of the theorem we shall construct polynomials $P_i(T)$ with the following properties:

$$0 \ne P_i(T) \in K_i[T] \quad , \quad P_i(\alpha_i) = 0 \quad , \quad H(P_i) \le (\Delta+1)^i H \quad ,$$

$$\text{Deg } P_i \le \Delta \text{ for } i \ge 0, \text{ and with Deg } P_i = 1 \text{ for } i>m.$$

If these polynomials have been so constructed, then since P_i is linear for $i>m$ we obtain $K_i = K_{i+1}$ and so $K_{m+1} = K'$. Furthermore, from the corollary above to Lemma 4 we obtain $H(\alpha_i) \le H(P_i) \le (\Delta+1)^i H$ for $i \ge 0$, and

$$G_{K_{i+1}} - G_{K_i} \le \frac{3}{2} H(P_i)(1 - 1/\Delta).$$

Hence

$$G_{K'} - G_K \le \frac{3}{2} H(1-1/\Delta) \sum_{i=0}^{m} (\Delta+1)^i \le \frac{3}{2}(\Delta+1)^m H$$

as required.

The construction of the polynomials P_i for $i \ge 0$ is achieved by induction. If we write $F(X,Y) = X^{r_0} F_0(X,Y)$, where F_0 is not divisible by

X, then the choice of $P_0(T) = F_0(0,T)$ has the required properties, so the case i=0 is established. We shall now assume that the polynomials $P_0,\ldots,P_{i-1}$ have already been constructed. Let us substitute $Y = \sum_{j=0}^{i-1} \alpha_j X^j + X^i Z$, so that $F(X,Y) = X^{r_i} F_i(X,Z)$, where F_i is some polynomial in X and Z not divisible by X. Each coefficient of F_i is a polynomial in $\alpha_0,\ldots,\alpha_{i-1}$ of degree at most Δ, so F_i has coefficients in K_i and

$$H(F_i) \le H(F) + \Delta \sum_{j=0}^{i-1} H(\alpha_j) \le (1+\Delta \sum_{j=0}^{i-1} (\Delta+1)^j)H = (\Delta+1)^i H$$

by the inductive hypothesis. Since $Z_0 = \sum_{j=i}^{\infty} \alpha_j X^{j-i}$ satisfies $F_i(X,Z_0) = 0$ we may choose $P_i(T) = F_i(0,T)$; P_i is non-zero, $\mathrm{Deg}\, P_i \le \Delta$ and $H(P_i) \le (\Delta+1)^i H$ as required.

The construction of the polynomials P_i, $i \ge 0$ is thus complete when we have established that P_i is linear for $i>m$. Now

$$X^{r_i} F_i(X,Z) = F(X,Y) = F(X, \sum_{j=0}^{i-1} \alpha_j X^j) + X^i Z \frac{\partial F}{\partial Y}(X, \sum_{j=0}^{i-1} \alpha_j X^j) + X^{2i} Z^2 \Phi(X,Z)$$

for some polynomial Φ in X and Z. Furthermore,

$$\frac{\partial F}{\partial Y}(X, \sum_{j=0}^{i-1} \alpha_j X^j) - \frac{\partial F}{\partial Y}(X, Y_0) = X^i \Theta(X)$$

for some power series Θ. Since $F_i(X,Z_0) = 0$ and $\frac{\partial F}{\partial Y}(X,Y_0)$ is divisible by X^m, we deduce that X^{m+i} divides $F(X, \sum_{j=0}^{i-1} \alpha_j X^j)$ for $i \ge m$, and so $r_i \ge i+m$ if $i \ge m$. However, we also have

$$X^i \frac{\partial F}{\partial Y}(X,Y_0) = X^{r_i} \frac{\partial F_i}{\partial Z}(X,Z_0) ;$$

since the left hand side is not divisible by X^{i+m+1} we have $r_i \le i+m$ for $i \ge 0$. Hence $r_i = i+m$ and $\frac{\partial F_i}{\partial Z}(0,\alpha_i)$ is non-zero for $i \ge m$. If $i>m$ then $2i>r_i$ and so $F_i(X,Z)$ may be written in the form $A_i(X) + ZB_i(X) + XZ^2 C_i(X,Z)$ with $B_i(0) \ne 0$; hence $P_i(T) = A_i(0) + TB_i(0)$, which is linear in T as required.

Finally we wish to establish the bound on $H(\alpha_j;\ m<j\le i)$ for each $i>m$. Let us write $\beta_j = \alpha_{j+m}$ for $j \ge 1$, so if $W = \sum_{j=1}^{\infty} \beta_j X^j$ then W satisfies $F_m(X,\alpha_m+W)=0$. Since $F_m(0,\alpha_m)=0$ and $\frac{\partial F_m}{\partial Z}(0,\alpha_m)\neq 0$, this equation contains no constant term, but a term δW for $\delta \neq 0$. Upon dividing by $-\delta$, we may therefore write $F_m(X,\alpha_m+W)=0$ in the form

$$W = \gamma_{10}X + \sum_{i+j\geq 2} \gamma_{ij}X^i W^j ;$$

let Γ denote the set of coefficients consisting of γ_{10} together with 1 and all the γ_{ij}, so

$$H(\Gamma) \leq H(F_m) + \Delta H(\alpha_m) \leq (\Delta+1)^{m+1} H.$$

We shall prove by induction on j that $v(\beta_j) \geq (2j-1)v(\Gamma)$ for each valuation v on L; it then follows that $H(\beta_j;1\leq j\leq i) \leq (2i-1)H(\Gamma)$ for each $i\geq 1$, as required. From the equation above for W we obtain $\beta_1=\gamma_{10}$, so $v(\beta_1)=v(\gamma_{10})\geq v(\Gamma)$ for each valuation v; let us make the inductive hypothesis $v(\beta_j) \geq (2j-1)v(\Gamma)$ for $1\leq j<i$. Now $W^j = \sum_{\ell=j}^{\infty} \beta_{j\ell}X^\ell$ for $j\geq 2$, where each $\beta_{j\ell}$ is given as the sum of terms $\beta_{\ell_1}\cdots\beta_{\ell_j}$ over indices $\ell_1\geq 1,\ldots,\ell_j\geq 1, \ell_1+\ldots+\ell_j=\ell$. Hence by the inductive hypothesis we deduce that $v(\beta_{j\ell}) \geq (2\ell-j)v(\Gamma)$ for $\ell \leq i+j-2$. Substituting in the equation for W above, and equating coefficients of X^i, we obtain

$$\beta_i = \gamma_{i0} + \sum_{\ell=0}^{i-1} \gamma_{\ell 1}\beta_{i-\ell} + \sum_{j=2}^{i} \sum_{\ell=0}^{i-j} \gamma_{\ell j}\beta_{j,i-\ell} ,$$

and so

$$v(\beta_i) \geq (2i-1)v(\Gamma)$$

as required. The proof of the lemma is thus complete.

The final part of this lemma may seem unnecessary, as we have already established the bound $H(\alpha_i) \leq (\Delta+1)^i H$ for all $i\geq 0$. If this estimate is employed for $i>m$, however, then the exponent of $\Delta+1$ in Theorems 9 and 12 would increase from degree 2 to degree 3, a substantial loss. We shall now extend the results in Lemma 5 to general Puiseux series, first over finite points in k, and secondly over ∞.

Lemma 6

Let $F(X,Y)$ *denote a polynomial in* X *and* Y *with coefficients in* K, *which is separable with respect to* Y. *Let us further denote by* α *an element of* K *and by* e *a positive integer. Suppose that if* $Y_o = \sum_{i=0}^{\infty} \alpha_i T^i$, *where each* α_i *lies in* k, *then* $F(T^e+\alpha, Y_o)$ *represents the zero power series in* T; *let* m *denote the order of vanishing in* T *of the series* $\frac{\partial F}{\partial Y}(T^e+\alpha, Y_o)$. *Then*

$$H(\alpha_i) \le (\Delta+1)^i (H+\Delta H(\alpha)) \qquad (0 \le i \le m),$$

$$H(\alpha_j; m<j\le i) \le (2i-2m-1)(\Delta+1)^{m+1}(H+\Delta H(\alpha)) \qquad (i>m) ,$$

where H *and* Δ *denote the height and total degree of* F *respectively. Furthermore, we have*

$$G_{K'} - G_K \le \frac{3}{2}(\Delta+1)^m (H+\Delta H(\alpha)) ,$$

where K' *is the field obtained from* K *by adjoining all the coefficients* α_i, $i \ge 0$.

Proof. If we write $G(T,Y) = F(T^e+\alpha, Y)$, then G is separable of degree at most Δ with respect to Y, and of height at most $H+\Delta H(\alpha)$. The results follow upon applying Lemma 5 to G.

Lemma 7

Let $F(X,Y)$ *denote a polynomial in* X *and* Y *with coefficients in* K, *which is separable with respect to* Y. *Let us further denote by* e *a positive integer. Suppose that if* $Y_o = \sum_{i=0}^{\infty} \alpha_i T^{i-e}$, *where each* α_i, $i \ge 0$, *lies in* k, *then* $F(T^{-e}, Y_o)$ *represents the zero Laurent series in* T. *Let* m *then denote the order of vanishing, which may be negative, of the Laurent series* $\frac{\partial F}{\partial Y}(T^{-e}, Y_o)$ *in* T, *and let* n *denote the integer* $m+(\Delta-1)e$. *Then* n *is non-negative, and*

$$H(\alpha_i) \le (\Delta+1)^i H \qquad (0 \le i \le n) ,$$

$$H(\alpha_j; n<j\le i) \le (2i-2n-1)(\Delta+1)^{n+1} H \qquad (i>n) ,$$

where H *and* Δ *respectively denote the height and total degree of* F. *Furthermore, we have*

$$G_{K'} - G_K \le \frac{3}{2}(\Delta+1)^n H ,$$

where K' *is the field obtained by the adjunction of all the coefficients* α_i, $i\ge 0$, *to* K.

Proof. Let us write $G(T,Z) = T^{\Delta e}F(T^{-e},T^{-e}Z)$, so G is a polynomial in T and Z with coefficients in K, which is separable of degree at most Δ with respect to Z, and of height H. If $Z_0 = \sum_{i=0}^{\infty} \alpha_i T^i$, then $G(T,Z_0)$ represents the zero power series in T. Furthermore,

$$\frac{\partial G}{\partial Z}(T,Z) = T^{e(\Delta-1)} \frac{\partial F}{\partial Y}(T^{-e},T^{-e}Z) ,$$

and so $\frac{\partial G}{\partial Z}(T,Z_0)$ vanishes to order n in T. The results now follow upon applying Lemma 5 to G.

Lemmata 6 and 7 provide bounds on the heights of the coefficients in each Puiseux expansion and, in conjunction with Lemma 5, yield an algorithm for the actual determination of these coefficients. In the construction of the rational functions required for the solution of equations of genera 0 and 1, we employed Lemma 1 which, by the use of Puiseux series, reduced an infinite set of valuation prescriptions to a finite system of linear equations. It is thus necessary to derive bounds on the heights of solutions of linear equations, which we now proceed to do.

Lemma 8

Let u_{ij}, $1\le i\le m$, $1\le j\le n$, *denote elements of* K *such that* $H(u_{ij};1\le j\le n) \le H$ *for each* i. *Then there exists a basis for the solutions in* K *of the linear equations*

$$\sum_{j=1}^{n} u_{ij}x_j = 0 \qquad (1\le i\le m),$$

such that each basis element $x_1,\ldots,x_n$ *satisfies*

$$H(x_1,\ldots,x_n) \le rH ,$$

where r *is the rank of the* m×n *matrix* $\underline{U}$ *formed by the coefficients* u_{ij}.

Proof. It may be assumed that the first r rows of $\underline{U}$ are linearly independent; the remainder may then be discarded, and so we may assume that m=r. Furthermore, after rearranging the columns of $\underline{U}$ and $x_1,\ldots,x_n$ correspondingly, we may suppose that the first r columns are linearly independent, and so form an r×r invertible matrix which we denote by $\underline{V}$. Let us also denote by $\underline{y}$ and $\underline{z}$ the column matrices with entries $x_1,\ldots x_r$ and $x_{r+1},\ldots,x_n$ respectively. The equations $\sum_{j=1}^{n} u_{ij}x_j = 0$, $1\le i\le m$, may now be written in the contracted form

$$\underline{V}.\underline{y} + \underline{W}.\underline{z} = 0 ,$$

where $\underline{W}$ is the r×(n-r) matrix formed from the last n-r columns of $\underline{U}$. Since $\underline{V}$ is invertible, we may rewrite this equation as

$$\underline{y} = -\underline{V}^{-1}.\underline{W}.\underline{z} .$$

Hence the coefficients of $\underline{z}$, namely $x_{r+1},\ldots,x_n$, may assume any values, and then the coefficients of $\underline{y}$, namely $x_1,\ldots,x_r$, are determined. Let us choose $\underline{z} = -(\det \underline{V})\underline{t}$, where the coefficients of $\underline{t}$ lie in the ground field k: such choices of $\underline{t}$ will generate over K the full set of solutions as required. We now have $\underline{y} = (\operatorname{adj} \underline{V}).\underline{W}.\underline{t}$, where adj $\underline{V}$ denotes the adjugate matrix, consisting of the cofactors of $\underline{V}$. Hence

$$x_i = \det(\underline{V}_i) \quad , \quad 1\le i\le r ,$$

where $\underline{V}_i$ is the r×r matrix obtained from $\underline{V}$ by replacing the i-th column by the column matrix $\underline{W}.\underline{t}$. Since we have chosen

$$x_{i+r} = -(\det \underline{V})t_i \quad , \quad 1\le i\le n-r ,$$

where each t_i lies in k, we obtain

$$v(x_1,\ldots,x_n) \ge \sum_{i=1}^{r} \min\{v(u_{ij});1\le j\le n\}$$

for each valuation v on K, and so $H(x_1,\ldots,x_n)\le rH$ as required.

The result of Lemma 8 may be extended to inhomogeneous linear equations.

Corollary

Let u_i, u_{ij}, $1 \le i \le m$, $1 \le j \le n$, *denote elements of* K *such that* $H(u_i, u_{ij}; 1 \le j \le n) \le H$ *for each* i. *If the equations*

$$\sum_{j=1}^{n} u_{ij} x_j = u_i \qquad (1 \le i \le m)$$

possess a solution $x_1, \ldots, x_n$ *in* K, *then there is a solution satisfying*

$$H(x_1, \ldots, x_n) \le rH ,$$

where r *is the rank of the* $m \times n$ *matrix* $\underline{U}$ *formed by the coefficients* u_{ij}.

Proof. Let $\underline{T}$ denote the $m \times (n+1)$ matrix obtained by adjoining the u_i, $1 \le i \le m$, as a final column to $\underline{U}$. Then since the inhomogeneous equations possess a solution, $\underline{T}$ also has rank r. Hence by Lemma 8 there exists a solution in K of the equations

$$\sum_{j=1}^{n} u_{ij} y_j + u_i y_{n+1} = 0 \qquad (1 \le i \le m)$$

such that $y_{n+1} \neq 0$ and

$$H(y_1, \ldots, y_{n+1}) \le rH.$$

The result now follows on choosing $x_i = -y_i / y_{n+1}$ for $1 \le i \le n$, since

$$H(x_1, \ldots, x_n) = H(-x_1, \ldots, -x_n, 1) \le H(y_1, \ldots, y_{n+1}) \le rH$$

as required.

We have now completed the preparations necessary for the proofs of Theorems 9 and 12 on the heights of the integral solutions of equations of genera 0 and 1 respectively. In §2 we shall establish Theorem 9 by making a close analysis of the construction in §2 of Chapter IV, and in §3 we shall study the method in §3 of Chapter IV to prove Theorem 12.

2 *GENUS ZERO*

This section is devoted to establishing Theorem 9, which asserts an upper bound on the heights of all integral solutions of an arbitrary equation of genus zero. The method employed for the proof consists of an analysis of the construction of the algorithm derived in §2 of Chapter IV for the complete solution of the equation, together with an estimation at each stage of the heights of the parameters introduced and a bound on the genus of the field they generate. For clarity we shall accordingly separate the proof into the several steps.

(i) First we shall introduce some notation: L will denote a sufficiently large finite extension of K, which will in fact be specified later (v); all the heights will refer to L unless otherwise stated. Initially $F(X,Y)$ will denote a polynomial of total degree Δ, which is also monic of degree Δ with respect to Y, of height H and with coefficients in K. At the close of the proof we shall transfer to the general case which is the subject of Theorem 9. As usual F is taken to be irreducible over $k(X)$, and such that the associated extension K has genus 0 over k, with K possessing at least 3 infinite valuations; hence $\Delta \geq 3$. Our object is to determine a bound on the height $H_K(X,Y)$ whenever X, Y lie in $\mathcal{O}$ and satisfy $F(X,Y) = 0$.

(ii) The discriminant of F with respect to Y will be denoted by $D(X)$. We shall require bounds on the height and degree of D, which we now obtain. Since D has coefficients consisting of polynomials in the coefficients of F, of total degree $2\Delta-2$, D has coefficients in K and

$$H(D) \leq (2\Delta-2)H.$$

This argument also yields $\mathrm{Deg}\, D \leq (2\Delta-2)\Delta$, but this bound may be improved in the present case, sufficient to affect the final result in Theorem 9. If L denotes the splitting field of F over $k(X)$, so that F has zeros $Y_1(=Y), Y_2, \ldots, Y_\Delta$ in L, then D is equal to the $\Delta\times\Delta$ Vandermonde determinant with entries Y_j^{i-1}, $1\leq i,j\leq\Delta$. Now let w denote an infinite valuation on L, so $w(X) < 0$. Since F has total degree Δ, the equation obtained for Y/X by dividing by X^Δ contains no positive powers of X, so each Y_j/X satisfies a monic equation whose coefficients are regular at w; hence $w(Y_j/X) \geq 0$ and $w(Y_j) \geq w(X)$ for $w|\infty$. Thus each element f of the i-th row of the matrix representing D satisfies $w(f) \geq (i-1)w(X)$, and so $w(D) \geq \frac{1}{2}\Delta(\Delta-1)$. Since $w(D) = \text{Deg}\, Dw(X)$ and $w(X) < 0$ we obtain

$$\text{Deg}\, D \leq \tfrac{1}{2}\Delta(\Delta-1).$$

(iii) We denote by K_1 the splitting field of D over K. In view of the corollary to Lemma 4 we deduce that if $D(\alpha) = 0$ then $H(\alpha) \leq 2\Delta H$, and that

$$G_{K_1} - G_K \leq \frac{3}{2}\Delta^3 H ,$$

since D has at most $\frac{1}{2}\Delta^2$ zeros.

(iv) The next stage is to adjoin to K_1 the coefficients of all the Puiseux expansions of Y at the zeros of D. If α is such a zero and ν is a valuation on K such that $\nu(X-\alpha) > 0$, then we denote the expansion of Y at ν by $\sum_{i=0}^{\infty} \alpha_{i\nu} X_\nu^i$, where $X_\nu = (X-\alpha)^{1/e_\nu}$. Now D is equal to the product over the conjugates of Y of the formal derivative $\frac{\partial F}{\partial Y}$, so for each α we obtain $\sum_{\nu|\alpha} \nu(\frac{\partial F}{\partial Y}) = \text{ord}_\alpha D$, and hence $\nu(\frac{\partial F}{\partial Y}) \leq \frac{1}{2}\Delta^2$ for each such ν. By Lemma 6 we now have

$$H(\alpha_{i\nu}) \leq 2(\Delta+1)^{i+2} H \qquad (0 \leq i \leq \tfrac{1}{2}\Delta^2) ,$$

$$H(\alpha_{j\nu}; \tfrac{1}{2}\Delta^2 < j \leq i) \leq 2(2i-\Delta^2-1)(\Delta+1)^{\frac{1}{2}\Delta^2+3} H \qquad (i > \tfrac{1}{2}\Delta^2) ,$$

since $H + \Delta H(\alpha) \leq 2(\Delta+1)^2 H$. Furthermore, if K_2 denotes the field obtained from K_1 by adjoining all these Puiseux coefficients, then Lemma 6 further yields

$$G_{K_2} - G_{K_1} \leq \frac{3}{2}\Delta^3 (\Delta+1)^{\frac{1}{2}\Delta^2+2} H ,$$

since there are at most $\frac{1}{2}\Delta^3$ expansions with coefficients to adjoin.

(v) Finally we form L by adjoining to K_2 all the coefficients of the Puiseux expansions of Y at ∞. If $\nu|\infty$ then since $\nu(Y) \geq \nu(X)$, the expansion of Y/X at ν may now be written as $\sum_{i=0}^{\infty} \alpha_{i\nu} X_\nu^i$, where $X_\nu = X^{-1/e_\nu}$.

As in Lemma 7 we denote by n_ν the integer

$$\nu(\frac{\partial F}{\partial Y}) + (\Delta-1)e_\nu .$$

We wish to bound n_ν, and from Lemma 7 $n_\nu \geq 0$ for each $\nu|\infty$. However, $\sum_{\nu|\infty} \nu(\frac{\partial F}{\partial Y}) = -\mathrm{Deg}\, D < 0$, so $\sum_{\nu|\infty} n_\nu \leq \Delta \sum_{\nu|\infty} e_\nu = \Delta^2$; hence $n_\nu \leq \Delta^2$. Lemma 7 now yields

$$H(\alpha_{i\nu}) \leq (\Delta+1)^i H \qquad (0 \leq i \leq \Delta^2) ,$$

$$H(\alpha_{j\nu}; \Delta^2 < j \leq i) \leq (2i-2\Delta^2-1)(\Delta+1)^{\Delta^2+1} H \qquad (\Delta^2 < i)$$

and

$$G_L - G_{K_2} \leq \frac{3}{2}\Delta\,(\Delta+1)^{\Delta^2} H ,$$

since there are at most Δ expansions whose coefficients are to be adjoined. Combining this last inequality with the bounds above on $G_{K_2} - G_{K_1}$ and $G_{K_1} - G_K$, we deduce that

$$G_L - G_K \leq 2(\Delta+1)^{\Delta^2+1} H. \tag{28}$$

(vi) The results preliminary to the construction have now been achieved; it is our next objective to determine an element T in K such that $\nu(T) \geq 0$ for $\nu \neq u$ and $u(T) = -1$, where u is some chosen infinite valuation on K. Any element T of K may be expressed uniquely in the form

$$TD(X) = \sum_{i=1}^{\Delta} A_i(X) Y^{i-1} , \tag{29}$$

where $A_1,\dots,A_\Delta$ are elements of $k(X)$, that is, rational functions of X with coefficients in k. If we denote by $T_1(=T),\dots,T_\Delta$ the functions in L corresponding to the values of T given by (29) as Y ranges over the conjugates $Y_1,\dots,Y_\Delta$ respectively, then we may invert the linear equations thus supplied and obtain

$$A_\ell = \det \underline{A}_\ell \qquad (1 \leq \ell \leq \Delta),$$

where $\underline{A}_\ell$ is the $\Delta\times\Delta$ matrix whose (i,j)-th coefficient is Y_j^{i-1} for $i\neq\ell$, and T_j for $i=\ell$. If w is a finite valuation on L, then since $w(Y_j)\geq 0$ and $w(T_j)\geq 0$ for each j we obtain $w(A_\ell)\geq 0$ for $w\nmid\infty$, $1\leq\ell\leq\Delta$; hence each A_ℓ is in fact a polynomial in X. Furthermore, since $w(Y_j)\geq w(X)$ and $w(T_j)\geq w(X)$ for $w|\infty$, then as for the Vandermonde determinant for D we obtain $w(A_\ell)\geq w(X)\{1+\sum_{i\neq\ell}(i-1)\}$ for $1\leq\ell\leq\Delta$, and so

$$\text{Deg } A_\ell \leq \tfrac{1}{2}\Delta^2 \qquad (1\leq\ell\leq\Delta).$$

We conclude that TD(X) may be expressed in the form

$$Z = \sum_{i=1}^{\Delta}\sum_{j=0}^{\frac{1}{2}\Delta^2}\beta_{ij}X^jY^{i-1}$$

for some elements β_{ij} in k, to be determined. As in the proof of Lemma 1 we may replace the infinite set of inequalities above which define T by the finite set

$$v(Z)\geq v(D) \qquad u\neq v|\infty \text{ or } v(D)>0, \qquad u(Z)\geq u(D)-1\ ;$$

the linear space thus defined for Z will have dimension 2: evidently $Z=D$ is a solution, and any choice of Z with $T=Z/D$ not in k will suffice for T as the uniformising parameter required. Now, by considering the Puiseux expansions of Z at the valuations concerned, we deduce that this set of inequalities is equivalent to a finite system of linear equations in the β_{ij}. Evidently the coefficients of each equation lie in L, and the rank of the system is $\frac{1}{2}\Delta(\Delta^2+2)-2$. We conclude from Lemma 8 that each β_{ij} may be chosen in L with

$$H(\beta_{ij};1\leq i\leq\Delta,\ 0\leq j\leq\tfrac{1}{2}\Delta^2)\leq\tfrac{1}{2}(\Delta+1)^3\,H_1\ ,$$

where H_1 is the maximum of the heights of the equations concerned. The next two steps will provide a bound on H_1.

(vii) If $v\nmid\infty$ and $v(D)>0$, then $v(D)\leq e_v\,\text{Deg } D\leq\frac{1}{2}\Delta^3$. Now $v|\alpha$ for some zero α of D, and so in the linear equations corresponding to the inequality $v(Z)\geq v(D)$, the coefficients α_{iv} occur for $i<v(D)$ with total

degree $\Delta-1$ and the coefficient α to degree $\frac{1}{2}\Delta^2$. Using the estimates in (iii) and (iv) on the heights of these coefficients, we deduce that the height of each equation does not exceed

$$(\Delta-1)\sum_{i=0}^{\frac{1}{2}\Delta^2} H(\alpha_{i\nu}) + (\Delta-1)H(\alpha_{j\nu};\tfrac{1}{2}\Delta^2<j<\tfrac{1}{2}\Delta^3) + \tfrac{1}{2}\Delta^2 H(\alpha) \leq 2(\Delta+1)^{\frac{1}{2}\Delta^2+7}H.$$

(viii) If $\nu|\infty$ then $\nu(D)<0$ and $\nu(X)=-e_\nu$. In the linear equations corresponding to $\nu(Z)\geq\nu(D)$ for $\nu\neq u$, or $u(Z)\geq u(D)-1$, the coefficients $\alpha_{i\nu}$ occur only for $i<e_\nu(\frac{1}{2}\Delta^2+\Delta-1)$, and with total degree $\Delta-1$. Since $e_\nu\leq\Delta$ the height of each equation does not exceed

$$(\Delta-1)\sum_{i=0}^{\Delta^2} H(\alpha_{i\nu}) + (\Delta-1)H(\alpha_{i\nu};\Delta^2<i<\Delta(\tfrac{1}{2}\Delta^2+\Delta-1)) \leq (\Delta+1)^{\Delta^2+5}H.$$

Since $\Delta\geq 3$, $H_1\leq(\Delta+1)^{\Delta^2+5}H$, and so

$$H(\beta_{ij};1\leq i\leq\Delta,0\leq j\leq\tfrac{1}{2}\Delta^2) \leq \tfrac{1}{2}(\Delta+1)^{\Delta^2+8}H.$$

(ix) Having thus determined the uniformising parameter T, we wish to compute the coefficients of the polynomials C and Q such that $X=C(T)/Q(T)$. As already noted in §2 of Chapter II, Q may in fact be taken as the product $\prod(T-T_\nu)^{e_\nu}$ over all the valuations $\nu\neq u$ on K with $\nu|\infty$, where T_ν lies in k such that $\nu(T-T_\nu)>0$. Thus T_ν is just the constant coefficient of the Puiseux expansion of T at ν, and so as above we deduce that T_ν lies in L and

$$H(T_\nu) \leq (\tfrac{1}{2}\Delta^2+\Delta)H(D)+H(\beta_{ij};i,j)+(\Delta+1)^{\Delta^2+5}H \leq (\Delta+1)^{\Delta^2+8}H,$$

since the coefficients of D occur in the expansion of $1/D$ in powers of $1/X$ with total degree at most $\frac{1}{2}\Delta^2+\Delta$. Hence

$$H(Q) \leq (\Delta+1)^{\Delta^2+9}H.$$

We now observe that, regardless of the polynomial C, we have $\nu(XQ(T)-C(T))\geq 0$ for all valuations $\nu\neq u$, both finite and infinite. It therefore suffices for the determination of C to solve the linear equations implied by the inequality $u(XQ(T)-C(T))>0$; by the sum formula

$XQ(T)-C(T)$ must then vanish identically. Now C has degree Δ, so we may write $C(T) = \sum_{i=0}^{\Delta} \gamma_i T^i$ for some $\gamma_0,\ldots,\gamma_\Delta$ in L, to be determined. The corollary to Lemma 8 in fact yields

$$H(\gamma_0,\ldots,\gamma_\Delta) \le (\Delta+1)\{H(Q) + \Delta H(S)\} ,$$

where S is the set of coefficients δ_i, $0\le i\le\Delta$, the Puiseux expansion of T at u being $\sum_{i=0}^{\infty} \delta_i x_u^{i-1}$. As for the estimation on the height of T_v above we obtain

$$H(S) \le (\Delta+1)^{\Delta^2+8} H ,$$

and so

$$H(C) = H(\gamma_0,\ldots,\gamma_\Delta) \le 2(\Delta+1)^{\Delta^2+10} H .$$

Since $v(YQ(T)) \ge 0$ for all valuations $v \ne u$, there exists a polynomial $B(T)$ with $YQ(T) = B(T)$. By considering the equations implied by the inequality $u(YQ(T) - B(T)) > 0$ as above we obtain

$$H(B) \le (\Delta+1)\{H(Q) + \Delta H(S) + \sum_{i=0}^{\Delta} H(\alpha_{iu})\} ,$$

and hence

$$H(B) \le 2(\Delta+1)^{\Delta^2+10} H .$$

(x) In order to determine a bound on $H(X,Y)$ from the information already obtained the following lemma is required.

Lemma 9

Suppose that $C(T)$ *and* $Q(T)$ *are coprime polynomials with coefficients in* L, *such that* $\operatorname{Deg} C = \Delta$, $\operatorname{Deg} Q < \Delta$, *and* Q *has at least* 2 *distinct zeros* β *and* γ *in* L. *Then whenever* T *is an element of* L *with* $Q(T) \ne 0$ *such that* $C(T)/Q(T)$ *lies in* O_L, *we have*

$$H(T) \le 9H + 2g_L + r_L - 1 ,$$

where g_L *is the genus of* L/k, r_L *is the number of infinite valuations on* L, *and* H *is equal to the largest of* $H(C)$, $H(Q)$, $\Delta H(\beta)$ *and* $\Delta H(\gamma)$.

Proof. The result will be obtained as a consequence of the analysis in Chapter IV, §2. We denote by $1/\alpha$ the leading coefficient of C, so that $P = \alpha C$ is monic, and $H(\alpha) \le H$, $H(P) \le H$. Let us denote by V the set of valuations v on L at which one or more of the following occur:

$$v|\infty,\ v(P) < 0,\ v(Q) < 0,\ v(\alpha) < 0,\ v(\beta-\gamma) > 0,\ v(P(\beta)) > 0,$$

$$v(P(\gamma)) > 0.$$

Evidently

$$|V| \le r_L + H(P) + H(Q) + H(\alpha) + H(\beta-\gamma) + H(P(\beta)) + H(P(\gamma)) ,$$

and so

$$|V| \le r_L + 8H ,$$

since $H(P(\beta)) \le 2H$, $H(P(\gamma)) \le 2H$, and $\Delta \ge 3$. As in §2 of Chapter IV, for each valuation v lying outside V we have $v(T-\beta) = v(T-\gamma) = v(\beta-\gamma) = 0$. The fundamental inequality (11), applied to the equation

$$(T-\beta) + (\gamma-T) + (\beta-\gamma) = 0$$

yields

$$H((T-\beta)/(\beta-\gamma)) \le |V| + 2g_L - 1 ,$$

since $|V| \ge 1$ and so the right hand side of this inequality is non-negative. We conclude that

$$H(T) \le 9H + 2g_L + r_L - 1$$

as required.

(xi) We now employ Lemma 9 to complete the proof of Theorem 9. For

$$2g_L + r_L - 1 - [L:K](2g + r - 1) \le 2(G_L - G_K) \le 4(\Delta+1)^{\Delta^2+1} H$$

by (28). Combining this with the inequalities already obtained on the heights of C and Q, we deduce that if T lies in L then

$$H(T) \le 19(\Delta+1)^{\Delta^2+10} H + [L:K](2g + r - 1).$$

Now, if X, Y is a solution in $\mathcal{O}$ of $F(X,Y) = 0$, then either $D(X) = 0$ or T lies in L, and since $X = C(T)/Q(T)$, $Y = B(T)/Q(T)$, in the latter case we have

$$H(X,Y) \le H(B(T),C(T)) + H(Q(T)) \le 40(\Delta+1)^{\Delta^2+11} H + 2\Delta[L:K](2g + r - 1).$$

If $D(X) = 0$, then $H(X) \le H(D) \le 2\Delta H$, and $H(Y) \le \Delta H(X) + H$, so $H(X,Y) \le 2(\Delta+1)^2 H$, and hence the above inequality on $H(X,Y)$ is actually valid for all X, Y in $\mathcal{O}$ with $F(X,Y) = 0$. However, we recall that X, Y and the coefficients of F lie in K, so dividing by $[L:K]$ we obtain

$$H_K(X,Y) \le 40(\Delta+1)^{\Delta^2+11} H_K(F) + 2\Delta(2g + r - 1)$$

as required.

(xii) The result obtained applies to a polynomial F which is monic with respect to Y; let us now consider the general polynomial F as in Theorem 9. As in §2 of Chapter II, the substitution $X = Z + \lambda Y$ for suitable choice of λ in k leads to a polynomial equation $G(Z,Y) = 0$, where G has height at most $H_K(F)$, total degree Δ, and is monic with respect to Y of degree Δ. Since Z lies in $\mathcal{O}$ whenever X and Y do, and $H_K(X,Y) = H_K(Z,Y)$, the result above enables us to conclude Theorem 9.

3 *GENUS ONE*

In this section we shall prove the bound in Theorem 12 on the heights of integral solutions of any equation of genus 1. The method of proof will consist of an analysis of the construction of the algorithm derived in §3 of Chapter IV for the complete solution of such an equation, together with a result (Lemma 10) on the heights of points on elliptic curves. As with the case of genus 0 in §2, we shall for clarity separate the proof into the several stages, many of which in fact may be transferred from §2 without alteration.

(i) We commence with the introduction of some notation: $F(X,Y)$ will denote a polynomial with coefficients in K, irreducible over the rational function field $k(X)$, and such that the associated extension K has genus 1 over k. By the argument of the concluding paragraph (xii) of §2, it suffices to establish Theorem 12 when F, of total degree Δ, is monic of degree Δ with respect to Y. We shall denote by L the field obtained from K by adjoining the zeros of the discriminant $D(X)$ of F, together with the coefficients of the Puiseux expansions of Y at these zeros and at ∞. All the heights employed will refer to L unless otherwise stated. The height of F will be denoted by H, so all the estimates in (i), (ii), (iii), (iv) and (v) of §2 apply without further comment.

(ii) The next stage is the construction of U and V in K such that $v(U) \geq 0$, $v(V) \geq 0$ for all valuations $v \neq u$, $u(U) = -2$, $u(V) = -3$, for some fixed infinite valuation u on K. As in (vi) of §2, we may express U and V in the form

$$UD(X) = \sum_{i=1}^{\Delta} U_i(X)Y^{i-1} \quad , \quad VD(X) = \sum_{i=1}^{\Delta} V_i(X)Y^{i-1} ,$$

where U_i, V_i, $1 \leq i \leq \Delta$, are polynomials in X with coefficients in k; furthermore, since $u(U) \geq 2u(X)$, $u(V) \geq 3u(X)$, we obtain

$$\operatorname{Deg} U_i \leq \tfrac{1}{2}\Delta^2+1 \quad , \quad \operatorname{Deg} V_i \leq \tfrac{1}{2}\Delta^2+2 \qquad (1 \leq i \leq \Delta),$$

since $\Delta \geq 3$. As before the coefficients of each U_i, V_i may be determined by the solution of the finite system of linear equations equivalent to the inequalities

$$v(U) \geq 0, v(V) \geq 0 \quad , \quad u \neq v|\infty \text{ or } v(D) > 0 \quad , \quad u(U) \geq -2, u(V) \geq -3.$$

Using the method of estimation in (vii), (viii), we deduce that the heights of these linear equations do not exceed $(\Delta+1)^{\Delta^2+5}H$, and so we may select the coefficients of the polynomials U_i and V_i in L, with

$$H(U_i,V_i;1\le i\le\Delta) \le (\Delta+1)^{\Delta^2+8}H .$$

We shall also require a bound on the heights of the initial Puiseux coefficients of U and V at u; if we denote their expansions at u by

$$\sum_{i=-2}^{\infty}\beta_i X_u^i \quad , \quad \sum_{i=-3}^{\infty}\gamma_i X_u^i$$

respectively, then each β_i and γ_i lies in L, and from the estimate on the heights of the Puiseux coefficients of Y at u we deduce that

$$H(S) \le (\tfrac{1}{2}\Delta^2+2\Delta+2)H(D) + H(U_i,V_i;1\le i\le\Delta) + (\Delta+1)^{\Delta^2+5}H$$

$$\le \frac{3}{2}(\Delta+1)^{\Delta^2+8}H,$$

where S is the set of coefficients β_i, γ_i with $i\le 2\Delta$. Now in §3 of Chapter IV we introduced the Puiseux coefficients of the expansions of U, V, U^2, UV, U^3, V^2 at u, these expansions being denoted by $\sum_{j=-6}^{\infty} h_{ij}X_u^j$, i=2,...,7, respectively. Now the coefficients h_{ij} with $j\le 0$ are each polynomials in the elements of S, of degree at most 3. Since these coefficients actually form the coefficients of the linear equations in $\delta_1,\dots,\delta_7$ such that

$$\delta_1 + \delta_2 U + \delta_3 V + \delta_4 U^2 + \delta_5 UV + \delta_6 U^3 + \delta_7 V^2 = 0, \qquad (31)$$

it follows that $\delta_1,\dots,\delta_7$ may be chosen in L, not all zero, with

$$H(\delta_1,\dots,\delta_7) \le 18H(S),$$

since the system of linear equations has rank 6.

(iii) We now proceed to determine polynomials Ξ, Ψ, Ω in $k[U]$ such that

$$X = \frac{\Xi(U) + V\Psi(U)}{\Omega(U)} .$$

As already observed in §3 of Chapter IV, Ω may be chosen to be the product $\prod(U-U_v)^{e_v}$, taken over all the valuations $v \neq u$ such that $v|\infty$, where U_v is the element of k satisfying $v(U-U_v) > 0$. Hence Deg $\Omega = \Delta - e_u$, and as in (ix) of §2, each element U_v lies in L satisfying

$$H(U_v) \leq \frac{3}{2}(\Delta+1)^{\Delta^2+8} H \qquad (u \neq v|\infty),$$

and so

$$H(\Omega) \leq \frac{3}{2}(\Delta+1)^{\Delta^2+9} H .$$

Now, regardless of the polynomials Ξ and Ψ we have $v(\Xi(U)+V\Psi(U)-X\Omega(U)) \geq 0$ for all valuations $v \neq u$ on K, both finite and infinite. It therefore suffices for the determination of Ξ and Ψ to solve the linear equations in their coefficients equivalent to the inequality

$$u(\Xi(U) + V\Psi(U) - X\Omega(U)) > 0.$$

We thus have

$$\max\{2 \text{ Deg } \Xi,\ 3 + 2 \text{ Deg } \Psi\} = 2\Delta - e_u,$$

and so the degrees of Ξ and Ψ are bounded. Since there are $2\Delta - e_u$ coefficients to be determined, the corollary to Lemma 8 yields

$$H(\Xi,\Psi) \leq (2\Delta-e_u)\{H(\Omega) + \Delta H(S)\},$$

since the coefficients of the equations involve the elements of S with total degree at most Δ. Using the estimates above for $H(\Omega)$ and $H(S)$ we obtain

$$H(\Xi,\Psi) \leq 6(\Delta+1)^{\Delta^2+10} H .$$

Now $v(Y) \geq v(X)$ for all $v|\infty$, so there exist polynomials $\Phi(U)$, $T(U)$ with

$$Y\Omega(U) = \Phi(U) + VT(U).$$

As above, it suffices for the determination of Φ and T to solve the linear equations in their coefficients equivalent to the inequality $u(\Phi(U)+VT(U)-Y\Omega(U))>0$. Hence

$$H(\Phi,T) \le (2\Delta-e_u)\{H(\Omega)+\Delta H(S)+\sum_{i=0}^{2\Delta} H(\alpha_{iu})\} \le 6(\Delta+1)^{\Delta^2+10}H .$$

(iv) Let us now replace V by $S=V+U\delta_5/2\delta_7+\delta_3/2\delta_7$. From the relation (31) we obtain

$$S^2 = \Theta(U) ,$$

where Θ is a cubic separable polynomial with coefficients in L, and

$$H(\Theta) \le 2H(\delta_1,\ldots,\delta_7) .$$

Furthermore, we also have

$$X\Omega(U) = \Lambda(U)+S\Psi(U) , \quad Y\Omega(U) = \Xi(U)+ST(U) ,$$

for some polynomials Λ and Ξ with coefficients in L. Finally, the estimates above yield

$$H(\Theta,\Omega,\Lambda,\Psi,\Xi,T) \le 16(\Delta+1)^{\Delta^2+10}H .$$

We shall determine a bound on $H(U)$ from the information already obtained: the necessary step is provided by the following lemma.

Lemma 10

Suppose that Θ is a cubic separable polynomial, that Ω, Λ, Ψ are polynomials whose degrees satisfy

$$\max\{2\ \mathrm{Deg}\ \Lambda,\ 3+2\ \mathrm{Deg}\ \Psi\} > 2\ \mathrm{Deg}\ \Omega,$$

and Θ, Ω, Λ, Ψ have coefficients in L with

$$H(\Theta,\Omega,\Lambda,\Psi) = H.$$

Then any solution U, S *in* L *of*

$$S^2 = \Theta(U)$$

such that $\Omega(U) \neq 0$ *and* $X = (\Lambda(U)+S\Psi(U))/\Omega(U)$ *is an element of* $\mathcal{O}_L$ *satisfies*

$$H(U) \leq 46H + 4(2g_L + r_L - 1),$$

where g_L *denotes the genus of* L/k *and* r_L *denotes the number of infinite valuations on* L.

Proof. Let us denote by L_1 the field obtained from L by adjoining the zeros of Θ, so Θ factorises in L_1 as

$$\Theta(U) = \theta(U-\theta_1)(U-\theta_2)(U-\theta_3) = \theta\Gamma(U)$$

say. We shall write

$$\delta = (\theta_2-\theta_3)^2(\theta_3-\theta_1)^2(\theta_1-\theta_2)^2,$$

so δ lies in L and satisfies $H(\delta) \leq 6H(\Gamma)$. Now let us denote by V the set of valuations v on L at which one or more of the following occur:

$$v|\infty,\ v(\Theta,\Omega,\Lambda,\Psi) < 0,\ v(\theta) > 0,\ v(\delta) > 0,\ v(\phi) > 0,$$

where ϕ is the leading coefficient of the polynomial $\Lambda^2-\Psi^2\Theta$. Hence

$$|V| \leq r_L + 11H,$$

since $H(\phi) \leq 3H$. However,

$$\Lambda^2(U) - \Psi^2(U)\Theta(U) - 2\Omega(U)\Lambda(U)X + \Omega^2(U)X^2 = 0$$

defines an equation for U, whose leading coefficient is ϕ by virtue of the assumption on the degrees of Λ, Ψ, Ω: we deduce that $v(U) \geq 0$ for v outside V.

Let us now adjoin to L_1 the square roots ξ_1,ξ_2,ξ_3 of $U-\theta_1, U-\theta_2, U-\theta_3$ respectively, where U, S is some solution in L of $S^2 = \Theta(U)$

such that X lies in $\mathcal{O}_L$; we shall denote by M the field generated over L_1 by ξ_1,ξ_2,ξ_3. As in the solution of the hyperelliptic equation in Chapter III, we shall derive a bound on H(U) from the identity

$$(\xi_2-\xi_3)+(\xi_3-\xi_1)+(\xi_1-\xi_2) = 0$$

and the three other identities relating $\xi_i \pm \xi_j$, $1\le i<j\le 3$. We first denote by $\mathcal{W}$ the set of valuations w on M such that $w|v$ for some v in $\mathcal{V}$. Now if w lies outside $\mathcal{W}$ then $w(U)\ge 0$, $w(\Gamma)\ge 0$, so $w(\theta_1,\theta_2,\theta_3)\ge 0$ and $w(\xi_1,\xi_2,\xi_3)\ge 0$. Since $(\xi_i-\xi_j)(\xi_i+\xi_j) = \theta_j-\theta_i$ and $w(\delta)=0$ for w outside $\mathcal{W}$ we obtain

$$w(\xi_i-\xi_j) = w(\xi_i+\xi_j) = 0 \qquad (1\le i<j\le 3, w\notin\mathcal{W}).$$

In the analysis of the hyperelliptic equation in §2 of Chapter III, the fundamental inequality (11), applied to the four equations relating $\xi_i\pm\xi_j$, $1\le i<j\le 3$, yields

$$H_M(U) \le 4(|\mathcal{W}| + 2g_M - 1) + 2H_M(\Gamma),$$

where g_M denotes the genus of M/k, and the heights refer to the field M. We seek to establish a bound on $|\mathcal{W}|+2g_M-1$. If w is any valuation on M, not necessarily in $\mathcal{W}$, then we denote by ε_w the ramification index of w over L. We shall prove in the next paragraph that $\varepsilon_w=1$ for w outside $\mathcal{W}$. This will suffice to complete the proof of the lemma, for the genus formula (4) yields

$$(2g_M-2) - [M:L](2g_L-2) = \sum_w(\varepsilon_w-1),$$

and so

$$2g_M + |\mathcal{W}| - 2 = [M:L](2g_L+|\mathcal{V}|-2),$$

since $\sum_{w|v}\varepsilon_w = [M:L]$ for each v. Dividing by $[M:L]$ we thus obtain

$$H(U) \le 4(2g_L+|\mathcal{V}|-1) + 2H(\Gamma),$$

which provides the required result.

It is now our object to prove that for each valuation w on M outside $\mathcal{W}$, we have $\varepsilon_w = 1$, that is, w is unramified over L. Now since $w(\theta) = 0$, $w(\theta_1, \theta_2, \theta_3, U) \geq 0$, and $w(\theta_i - \theta_j) = 0$ for $1 \leq i < j \leq 3$, the inequalities $w(U - \theta_i) > 0$, $w(U - \theta_j) > 0$ cannot hold simultaneously for $i \neq j$, so $w(U - \theta_i) = 0$ or $2w(S)$. Since S lies in L, we conclude that in either case w does not ramify over L_1. However, from the last sentence in the proof of Lemma 4 we now deduce that if w is ramified over L then either $w(\Gamma) < 0$ or $w(\delta) > 0$. Since neither of these occur for w outside $\mathcal{W}$, w is unramified over L and the proof of Lemma 10 is complete.

(v) We now wish to combine the result of Lemma 10 with the bounds already obtained to conclude Theorem 12. Now if X, Y is a solution in $\mathcal{O}$ of $F(X,Y) = 0$ with $D(X) \neq 0$, then U and S are elements of L satisfying $S^2 = \Theta(U)$. If in addition $\Omega(U) \neq 0$, then the expressions for X and Y in terms of U and S yield

$$H(X,Y) \leq H(\Lambda, \Psi, \Xi, T, \Theta) + H(\Omega) + 2\Delta H(U) ,$$

and so from Lemma 10 we obtain

$$H(X,Y) \leq 1472(\Delta+1)^{\Delta^2+11} H + 8\Delta(2g_L + r_L - 1) .$$

If $D(X) = 0$, then $H(X) \leq H(D) \leq 2\Delta H$ and $H(Y) \leq \Delta H(X) + H$, so $H(X,Y) \leq 2(\Delta+1)^2 H$. Let us now consider the remaining case, when $D(X) \neq 0$ but $\Omega(U) = 0$. Let us denote the values of X, Y and U at such a solution by x, y and u respectively, so $u = U_\upsilon$ for some $\mathcal{U} \neq \upsilon | \infty$. Now $\Theta(u) \neq 0$, for otherwise K has only one valuation υ at which $\upsilon(U - u) > 0$, and X has a pole at such a valuation since $\Omega(u) = 0$. We shall consider the identity

$$\Lambda^2(U) - \Psi^2(U)\Theta(U) - 2X\Omega(U)\Lambda(U) + X^2\Omega^2(U) = 0.$$

Since we may first extract any common factor of Λ, Ψ, Ω without increase of height, we may assume that $\Lambda(u) \neq 0$ or $\Psi(u) \neq 0$. Substituting $U = u$ we obtain $\Lambda^2(u) = \Psi^2(u)\Theta(u)$, so since $\Theta(u) \neq 0$ we necessarily have $\Lambda(u) \neq 0$. Hence

$$A(u) = 2x\Lambda(u)B(u) ,$$

where $A(U)$ and $B(U)$ are the polynomials obtained by dividing $\Lambda^2-\Psi^2\Theta$ and Ω by $(U-u)^n$, where $0<n=\mathrm{ord}_u\Omega$. Since $B(u)\neq 0$, we have $H(x)\leq H(A(u))+H(\Lambda(u)B(u))$, and thus

$$H(x,y)\leq 81(\Delta+1)^{\Delta^2+11}H.$$

Hence in both exceptional circumstances the bound obtained above on $H(X,Y)$ in the general case remains valid for all X, Y in $\mathcal{O}$ with $F(X,Y)=0$. Now

$$(2g_L+r_L-1)-[L:K](2g+r-1)\leq 2G_L-2G_K\leq 4(\Delta+1)^{\Delta^2+1}H$$

by (28). However, we recall that X, Y and the coefficients of F lie in K, so dividing by $[L:K]$ we obtain

$$H_K(X,Y)\leq 1473(\Delta+1)^{\Delta^2+11}H_K(F)+8\Delta(2g+r-1)$$

as required. The proof of Theorem 12 is thus complete.

CHAPTER VI FIELDS OF ARBITRARY CHARACTERISTIC

1 *INTRODUCTION*

Until now we have been concerned in this book only with equations over fields of characteristic zero. We shall now broaden our approach and provide complete solutions for the general Thue and hyperelliptic equations over fields of non-zero characteristic. In both cases we shall construct algorithms for the determination of all integral solutions of these equations, and establish criteria for the existence of infinitely many solutions. As in our work on the Thue and hyperelliptic equations in characteristic zero, the crux of the method is formed by a fundamental inequality, which must now apply to fields of arbitrary characteristic. The analysis actually reveals a much richer structure than for fields characteristic zero, and in order to explain why this arises, and to discuss the significance of the previous work on this subject, we shall make a brief survey of the foundations of transcendence theory by commencing with a discussion on the approximation of algebraic numbers by rationals, which began with the celebrated theorem of the French mathematician Liouville.

In 1844 Liouville *[19]* started the theory of transcendence by discovering the existence of a class of numbers which satisfied no algebraic equation over the integers. He in fact proved that any algebraic number cannot be too closely approximated by rationals; thus any number which is, such as $\sum_{n=1}^{\infty} 10^{-n!}$, must be transcendental. For if α is any algebraic number of degree $n(>1)$ over the rationals, then Liouville showed that there is a positive number c, depending only on α, such that

$$\left|\alpha - \frac{p}{q}\right| \geq \frac{c}{|q|^{\kappa}} \tag{32}$$

for all integers p, q with κ equal to n. This result is best possible for κ only if n is 2, and an important branch of transcendence is devoted to the reduction of the exponent κ in (32) for $n > 2$; c may then depend on both α and κ. It is immediate that the validity of (32) for some $\kappa < n$ when n is at least 3 yields Thue's theorem (see Chapter I); for any solution x, y of the Thue equation $F(x,y) = 1$ with x, y sufficiently large provides a too close approximation x/y to one of the roots of $F(\alpha,1) = 0$, so x and y are in fact bounded, as required. Accordingly Thue proved in 1909 that (32) holds for any $\kappa > \frac{1}{2}n + 1$; this result was improved by Roth *[33]* in 1955 to $\kappa > 2$, and this is essentially best possible (see *[7]*). It should be mentioned, though, that in the work of Thue and Roth no effective determination is made of the parameter c, depending on α and κ; hence Thue's work does not yield effective bounds on the integers x and y satisfying $F(x,y) = 1$. In contrast, for Liouville's theorem c may be computed readily, for if d denotes the denominator of α, then $d(p\alpha-q)$ is a non-zero algebraic integer of degree n, and so $d^n|N(p\alpha-q)| \geq 1$, where $N(x)$ is the product of the conjugates of x over $\mathbb{Q}$: Liouville's theorem follows.

We shall now consider the analogues of the theorems just discussed in the present context of function fields. Accordingly, let k denote an algebraically closed field, of arbitrary characteristic, and let $k(z)$ denote the rational function field over k. Any element of $k(z)$ may be expressed as a formal Laurent series in powers of $1/z$, and if we denote by $k\{1/z\}$ the field of all such formal Laurent series, then the canonical valuation ord f on $k\{1/z\}$ induces the infinite valuation on $k(z)$, as in Chapter I, §2. We shall employ briefly the multiplicative valuation $|f| = \exp(-\text{ord } f)$ for the purpose of direct comparison with the example of number fields. In exactly the same fashion as in the proof above of Liouville's theorem, we deduce that if α is an element of $k\{1/z\}$, algebraic of degree n (>1) over $k(z)$, then there exists a positive number c, depending only on α, such that

$$\left|\alpha - \frac{P}{Q}\right| \geq \frac{c}{|Q|^n} \tag{33}$$

for all polynomials P, Q in $k[z]$. By following Roth's method described above, Uchiyama *[44]* proved in 1959 that when k has characteristic zero the exponent n in (33) may be reduced to any $\kappa > 2$, with c now dependent on both α and κ. However, an example of Mahler *[20]* shows that no improvement on the exponent n in (33) is possible in general. For if k has characteristic p, then the Laurent series $\alpha = \sum_{m=0}^{\infty} z^{-p^m}$ satisfies $\alpha^p - \alpha + 1/z = 0$, and so is algebraic of degree p over k(z). Furthermore, if for each $t \geq 0$ we denote by Q_t and P_t the polynomials z^{p^t} and $z^{p^t} \sum_{m=0}^{t} z^{-p^m}$ respectively, then

$$|\alpha - \frac{P_t}{Q_t}| = |Q_t|^{-p},$$

and $|Q_t| = e^{p^t}$ is unbounded as t increases. It follows that (33) is indeed best possible with respect to n in this case. In 1968 Armitage claimed to have proved [1] that Uchiyama's theorem applies also to fields of positive characteristic, subject only to a simple restriction on the field extension obtained by adjoining α to k(z). Both assertion and "proof" were shown to be false by Osgood in 1975 [32], and he also demonstrated that an improvement on (33), to $\kappa = n-1$, is possible provided that α does not satisfy a Riccati differential equation.

Mahler's example provides an illustration of a richness and variety, previously absent when concerned only with fields of zero characteristic, and which we see necessarily reflected in the analysis of equations over fields of positive characteristic. For example, if X, Y is any integral solution of the Thue equation $X(X-Y)(X+Y) = 1$, over a field of characteristic p (>2), then so is X^p, Y^p. Since $H(X^p,Y^p) = pH(X,Y)$, the heights of solutions in $\mathcal{O}$ may be unbounded, in direct contrast to the case of characteristic zero (Theorem 3). In Chapter VII we shall in fact establish that it is possible to determine effectively a finite set $X_1,Y_1;\ldots;X_m,Y_m$ of solutions in $\mathcal{O}$ of $X(X-Y)(X+Y) = 1$ such that any non-constant solution in $\mathcal{O}$ is given by $X = X_i^{p^j}$, $Y = Y_i^{p^j}$ for some $1 \leq i \leq m$, $j \geq 0$.

We shall begin our analysis of equations over function fields of positive characteristic by a discussion of valuations, derivations and the genus, leading to a generalisation (Lemma 10) of the fundamental inequality. It is the result derived from this (Lemma 11), though, which assumes the greater importance in this context. In Chapter VII we shall employ Lemma 11 to discover algorithms whereby all the integral solutions of the general Thue and hyperelliptic equations may be determined. We observe that as before it is necessary for the actual computation of solutions to assume that the ground field k is presented explicitly. It is further necessary to assume that k is presented in such a way that, given any two elements a, b in k, we may determine effectively whether there exists a power q of the characteristic of k such that $a^q = b$, and, if so, such a q may be computed. The significance of this technical condition will become apparent in Chapter VII. We observe here that the requirement is satisfied if k is presented suitably as the algebraic closure of any finitely generated extension of a finite field. We see from the familiar principle of Lefschetz *[18]* that the information provided in any equation and field k may be so presented. The condition is thus not so much a restriction, but rather a matter of description.

2 PRELIMINARIES AND THE FUNDAMENTAL INEQUALITY

In this section we shall discuss the concepts of valuations, derivations and the genus as they appertain to function field of characteristic p; we shall conclude with a modification of the fundamental inequality applicable in the present context. Let us denote by K a finite extension of the rational function field k(z), where k is an algebraically closed field of characteristic $p > 0$; as before $\mathcal{O}$ denotes the integral closure of $k[z]$ in K. We begin by showing that it may be assumed that K is in fact separable over k(z). For if K/k(z) has inseparability degree p^n, then we shall prove the existence of Z in K with $Z^{p^n} = z$; thus K/k(Z) is separable, and f in K is integral over $k[Z]$ if and only if it is integral over $k[z]$ and so we may replace z by Z without loss. Now k is algebraically closed and so *a fortiori* perfect; it follows (see *[11]*) that there exists x in K with K/k(x) separable. Hence K is generated over k(x) by a single element, y say, and thus, as in the case of characteristic zero (see Chapter I, §2), the mapping $f \mapsto f'$ on k(x), namely differentiation with respect to x, extends uniquely to a derivation on K. However, if $f = h^p$ for some h in K, then $f' = 0$. We shall establish the converse of this result; moreover we shall show that the three conditions

$$t' \neq 0, \quad t \text{ is not a } p\text{-th power in } K, \quad K/k(t) \text{ is finite and separable} \qquad (34)$$

are equivalent for t in K. For we have already observed that if $t = s^p$ for some s in K, then $t' = 0$, and either t lies in k, in which case K/k(t) is infinite, or t does not lie in k, in which case $K/k(t)$ is finite but inseparable. Let us now suppose that t has minimal polynomial Q over $k(x)$, where

$$Q(T,x) = T^m + Q_1(x)T^{m-1}+\ldots+Q_m(x),$$

with $Q_1,\ldots,Q_m$ in $k(x)$. Differentiating, we obtain

$$0 = t'\frac{\partial Q}{\partial T}(t,x) + \frac{\partial Q}{\partial x}(t,x) ,$$

and $\frac{\partial Q}{\partial T}(t,x)$ is non-zero since $K/k(x)$ is separable; hence $t'=0$ if and only if $\frac{\partial Q}{\partial x}(t,x)=0$, that is, since Q is minimal, if and only if $Q_i'(x)=0$ for $i=1,\ldots,m$. By factorising a rational function f in $k(x)$ into linear factors, we deduce that $f'=0$ if and only if $f=h^p$ for some h in $k(x)$. We conclude that if $t'=0$ then $Q_i=P_i^p$ for some $P_1,\ldots,P_m$ in $k(x)$, and so

$$Q(T,x) = (S^m + P_1S^{m-1}+\ldots+P_m)^p \qquad \text{where } S^p=T;$$

thus $s=t^{1/p}$ has degree at most m over $k(x)$, and so lies in $k(x,t)$ as required. Finally, if $t'\neq 0$, then $Q(t,X)=0$ constitutes a separable equation for x over $k(t)$, and so $K/k(t)$ is separable; we have thus established the equivalence of the three conditions (34) on an element t of K. It follows from the equivalence of the last two that if $K/k(z)$ has inseparability degree p^n, then z is a p^n-th power in K, as required. We shall assume henceforth, as we may, that $K/k(z)$ is separable, generated by an element y of degree d, say, over $k(z)$. We shall denote by f' the derivative with respect to z of an element f in K. The result just established, that f' vanishes precisely when f is a p-th power in K, will play an important role, both in Lemmata 10 and 11 below, and in the subsequent analyses in Chapter VII of the Thue and hyperelliptic equations.

We shall now consider the canonical valuations on $k(z)$ and the construction of valuations on K (see Chapter I, §2). If a is any finite point of k, then we may embed $k(z)$ in the local field $k\{z-a\}$ of formal Laurent expansions in powers of $z-a$. The canonical valuation ord_a on $k\{z-a\}$ induces a valuation on $k(z)$; similarly $k(z)$ may be embedded in $k\{1/z\}$ and we thereby obtain the valuation ord. We wish to construct similar valuations on K. Let us denote by P the minimal polynomial of y

over $k(z)$. Although P is irreducible over $k(z)$, it may factorise when considered over the larger field $k\{z-a\}$; let us denote the factorisation by $P_1 \ldots P_{r_a}$, where P_i is irreducible over $k\{z-a\}$ of degree e_{ai}, say: the factors are all distinct and separable since P is itself separable. For each i, $1 \le i \le r_a$, we may adjoin a zero y_i of P_i to $k\{z-a\}$: the extension so obtained is *totally ramified*, and hence (see [17]) there exists a z_v such that the extension takes the form $k\{z_v\}$, the field of formal Laurent expansions in powers of z_v, and we now write $e_v = e_{ai}$ for the degree of $k\{z_v\}$ over $k\{z-a\}$. In the case when e_v is not divisible by p, the extension is said to be *tamely ramified* and we may choose $z_v = (z-a)^{1/e_v}$ as in the case of characteristic zero; however, when p does divide e_v this choice is impossible since $k\{z_v\}$ is always separable over $k\{z-a\}$. For all e_v the mapping $\sigma_v : y \mapsto y_i$ extends uniquely to give an embedding of K in the field $k\{z_v\}$, and we thereby obtain a valuation v on K, where $v(f)$, for f non-zero, is the order of vanishing of the Laurent series $\sigma_v(f)$ in powers of z_v. We also write $v|a$, since $v(f) = e_v \mathrm{ord}_a f$ for f in $k(z)$; such valuations v, for any a in k, are termed *finite*. Arguing similarly for the embedding of $k(z)$ in $k\{1/z\}$, we obtain the *infinite* valuations $v_1, \ldots, v_r$ on K, and we write $v_i|\infty$ for $1 \le i \le r$. For each non-zero element f in K *the sum formula*

$$\sum_v v(f) = 0 \tag{1}$$

is readily established as before. We also define the height $H(f) = -\sum_v \min(0, v(f))$, as in Chapter I.

On each of the local fields $k\{z_v\}$ we may define a derivation $\frac{d}{dv}$, namely differentiation with respect to z_v; for f in K, the order of vanishing $v(\frac{df}{dv})$ of this derivative is actually independent of the local parameter chosen at v. Furthermore, since $\frac{df}{dv} = f'\frac{dz}{dv}$, the sum $\sum_v v(\frac{df}{dv})$ is, by the sum formula, independent of f such that $f' \neq 0$. We may thus define the *genus* g of K by

$$2g - 2 = \sum_v v(\frac{df}{dv}) \qquad (f \notin K^p). \tag{5}$$

Finally the inequalities (6) apply also in the case of characteristic p, so we may restate our fundamental inequality (Lemma 2); the proof in Chapter I, §3 remains valid.

Lemma 10

Suppose that γ_1, γ_2 and γ_3 are non-zero elements of K *with* $\gamma_1+\gamma_2+\gamma_3=0$, *and such that* $v(\gamma_1)=v(\gamma_2)=v(\gamma_3)$ *for each valuation* v *not in the finite set* V. *Then either* γ_1/γ_2 *is a p-th power in* K, *or*

$$H(\gamma_1/\gamma_2) \le |V| + 2g - 2 . \tag{11}$$

In contrast with the case of characteristic zero, it is the first alternative in the lemma which presents the interesting case: it is quite feasible for γ_1, γ_2, γ_3 to be, say, p^m-th powers of elements δ_1, δ_2, δ_3 in K; the hypotheses of the lemma apply to δ_1, δ_2, δ_3, yet $H(\gamma_1/\gamma_2)=p^mH(\delta_1/\delta_2)$, which may be arbitrarily large. Let us now consider how this relates to the Thue equation

$$(X-\alpha_1Y)\dots(X-\alpha_nY) = \mu, \tag{9}$$

where $\alpha_1,\dots,\alpha_n,\mu$ lie in O, μ is non-zero and α_1, α_2, α_3 are distinct. Just as for characteristic zero in Chapter II, we write $\beta_i = X-\alpha_iY$, $1\le i\le n$, so if X, Y lie in O so do $\beta_1,\dots,\beta_n$; $\beta_1\dots\beta_n=\mu$ and

$$\beta_i(\alpha_j-\alpha_\ell) + \beta_j(\alpha_\ell-\alpha_i) + \beta_\ell(\alpha_i-\alpha_j) = 0 \tag{10}$$

for any three suffixes i, j, ℓ. As before it is this identity which forms the first step in our solution, and here we simplify it further to

$$\alpha+\beta = 1, \tag{35}$$

where $\alpha=\beta_i(\alpha_j-\alpha_\ell)/\beta_j(\alpha_i-\alpha_\ell)$, $\beta=\beta_\ell(\alpha_i-\alpha_j)/\beta_j(\alpha_i-\alpha_\ell)$, and α_i, α_j, α_ℓ are assumed distinct. We observe in connexion with (35) that if ν denotes the element $\mu(\alpha_j-\alpha_\ell)(\alpha_\ell-\alpha_i)(\alpha_i-\alpha_j)$ of O, then each of $\nu\alpha$, ν/α and ν/β lie in O. We shall show in Lemma 11 that these conditions are sufficient to determine the complete family of solutions to (35). As above, we note that these solutions may have unbounded heights, for if η and $1-\eta$ are both units in O, then $\alpha=\eta^q$, $\beta=(1-\eta)^q$ represents a solution of (35) for any power q of p, and $\nu\alpha$, ν/α, ν/β lie in O, but $H(\alpha)=qH(\eta)$, which becomes arbitrarily large as q increases, for η outside k. We shall show in Lemma 11 that this is essentially the only exceptional case; we shall also prove that only finitely many units η exist such that $1-\eta$ is also a unit and η is not a p-th power in K: throughout we denote these units by $\eta_1,\dots,\eta_s$.

Lemma 11

Suppose that ν *is a fixed non-zero element of* $\mathcal{O}$, *and* $\alpha \neq 0, 1$ *lies in* K, *subject to the conditions that* $\nu\alpha$, ν/α *and* $\nu/(1-\alpha)$ *lie in* $\mathcal{O}$. *If* α *is neither in* k, *nor equal to some power* p^j $(j \geq 0)$ *of one of the units* $\eta_1,\dots,\eta_s$, *then* α *has only finitely many possibilities. Furthermore, these possibilities, and the units* $\eta_1,\dots,\eta_s$ *defined above, all may be determined effectively.*

Proof. Let us first assume that α is not a p-th power in K; we shall prove that the possibilities for α are then but finite in number and effectively determinable, and they include the units $\eta_1,\dots,\eta_s$. Now if v is a finite valuation on K, then

$$-v(\nu) \leq v(\alpha) \leq v(\nu) ,$$

since $\nu\alpha$ and ν/α lie in $\mathcal{O}$; the same inequalities obtain also for $v(1-\alpha)$. Hence if we denote by V the set of valuations v on K such that $v|\infty$ or $v(\nu) > 0$, then $v(\alpha) = v(1-\alpha) = 0$ for v outside V. Lemma 10 may now be applied to the equation $\alpha + (1-\alpha) + (-1) = 0$, and we obtain $H(\alpha) \leq |V| + 2g - 2$. However,

$$H(\alpha) \quad = \quad -\sum_{v} \min(0,v(\alpha)) \quad = \quad \sum_{v} \max(0,v(\alpha)),$$

and so the complete set of values assumed by α has only finitely many possibilities. The proof of Lemma 1 applies to non-zero characteristic, so we conclude that α has only finitely many possibilities, effectively determinable, up to a factor in k. We may thus write $\alpha = a\gamma$, where a lies in k and γ belongs to some finite computable subset of K. The same conclusion also applies to $1-\alpha$, and so we may write $1-\alpha = b\delta$, where b lies in k and δ belongs to the same computable subset of K as does γ. Since α is not a p-th power in K by assumption we have $\gamma'\delta' \neq 0$. But $a\gamma + b\delta = 1$, and thus, upon differentiating, we obtain $a\gamma' + b\delta' = 0$, and so a and b are determined uniquely. We conclude that if α is not a p-th power in K then it has only a finite number of effectively determinable possibilities, as required.

We now turn to the general case, where we wish to determine all the possibilities for α in K. We may assume that α is not in k, and so $K/k(\alpha)$ is finite, with inseparability degree a power of p, q say, possibly equal to 1. It follows from the equivalence of the last two conditions (34) that $\alpha = \varepsilon^q$ for some ε in K, not a p-th power. It therefore suffices to determine all the possibilities for ε and q. Now $1-\alpha = (1-\varepsilon)^q$, so

$$|v(\varepsilon)| \le v(\nu)/q, \qquad |v(1-\varepsilon)| \le v(\nu)/q$$

whenever v is a finite valuation on K. Since ε is not a p-th power in K, we conclude as above that it has only finitely many possibilities. The proof of the lemma is completed by showing that either ε and $1-\varepsilon$ are units in $\mathcal{O}$, or q is bounded. But if ε and $1-\varepsilon$ are not both units then from the two inequalities above we obtain $v(\nu) \ge q$ for some $v \nmid \infty$, so

$$q \le \max\{v(\nu) \ ; \ v \nmid \infty\} \le H(\nu) \ ,$$

and thus q is bounded as required.

CHAPTER VII SOLUTIONS FOR NON-ZERO CHARACTERISTIC

1 THE THUE EQUATION

In this section we shall solve completely the general Thue equation over an arbitrary algebraic function field of positive characteristic. It is Lemma 11 which plays the crucial role, both in this analysis and in the subsequent resolution in §3 of the hyperelliptic equation. The algorithm for the Thue equation will be illustrated in §2 by a simple example. Let us denote by $\alpha_1,\ldots,\alpha_n,\mu$ elements of $\mathcal{O}$ such that $\alpha_1,\alpha_2,\alpha_3$ are distinct and μ is non-zero; we shall establish the following theorem.

Theorem 13

All the solutions X, Y *in* $\mathcal{O}$ *of the Thue equation*

$$(X - \alpha_1 Y)\ldots(X - \alpha_n Y) = \mu \tag{9}$$

may be determined effectively.

In fact we shall describe the various infinite families of solutions and determine precisely when they arise. In particular, it may be determined effectively when (9) has only finitely many solutions in $\mathcal{O}$. However, in contrast with the case of characteristic zero (Theorem 2), it is not possible to state a simple criterion for the existence of only finitely many solutions in $\mathcal{O}$. For example, if $p>3$ and η, $\eta-\xi$, $\eta+\xi$ are units in $\mathcal{O}$ such that $\eta(\eta-\xi)(\eta+\xi) = \mu$ is not a cube in $\mathcal{O}$, then the equation $X(X-Y)(X+Y) = \mu$ cannot be transformed by a linear substitution, with coefficients in $\mathcal{O}$, into an equation over k, but it does have infinitely many solutions given by $X = \eta^q \mu^{(1-q)/3}$, $Y = \xi^q \mu^{(1-q)/3}$, where q is any power of p such that 3 divides q-1. To pursue this point further, we see that the equation $X(X-Y)(X+Y) = \mu$ may be transformed by the substitution $X = x\mu^{1/3}$, $Y = y\mu^{1/3}$ into $x(x-y)(x+y) = 1$, which has coefficients in k as required; but now the substitution has coefficients in $K(\mu^{1/3})$, not in K. We shall see that the existence of infinitely many solutions in $\mathcal{O}$ of (9) does imply that (9) may be transformed by a linear substitution into an equation over k, with the coefficients of the substitution in some radical extension $K(\phi^{1/n})$ of K. However, we shall also see that this condition, whilst being necessary, is not sufficient, and that the precise condition involves considerably more analysis. Nevertheless, this example does provide the typical form of infinite family of solutions arising. We shall actually prove that, apart from a finite number of such families and the infinite family of bounded height already found in characteristic zero, there are only finitely many solutions in $\mathcal{O}$ of (9); furthermore, these solutions, and the infinite families may all be determined effectively when they exist.

Let X, Y denote a solution in $\mathcal{O}$ of the Thue equation (9); as before we shall write, for brevity, $\beta_i = X - \alpha_i Y$, $1 \le i \le n$. Then $\beta_1, \ldots, \beta_n$ are elements of $\mathcal{O}$ such that $\beta_1 \ldots \beta_n = \mu$ and, for any three suffixes i, j, ℓ, we have the identity

$$\beta_i(\alpha_j - \alpha_\ell) + \beta_j(\alpha_\ell - \alpha_i) + \beta_\ell(\alpha_i - \alpha_j) = 0. \tag{10}$$

Provided $\alpha_i, \alpha_j, \alpha_\ell$ are distinct, we may apply Lemma 11 with $\alpha = \beta_i(\alpha_j - \alpha_\ell)/\beta_j(\alpha_i - \alpha_\ell)$ and $\nu = \mu(\alpha_j - \alpha_\ell)(\alpha_\ell - \alpha_i)(\alpha_i - \alpha_j)$. We conclude that α is either an element of the ground field k, or equal to some unit η_r^q, $1 \le r \le s$, $q = p^t$, $t \ge 0$, or a member of some finite computable subset of K. By considering the various values of the suffix ℓ, we shall further limit the range of possibilities for α. Let us first observe that any ratio β_j/β_i for $\alpha_i \neq \alpha_j$ determines X and Y up to an n-th root of unity in k. For by (10) β_j/β_i determines all the ratios β_ℓ/β_i, $1 \le \ell \le n$, and so β_i is determined up to a factor of an n-th root of unity by the equation $\beta_1 \ldots \beta_n = \mu$. We may now determine X and Y from the equations

$$Y = \frac{\beta_j - \beta_i}{\alpha_i - \alpha_j}, \qquad X = \beta_i + \alpha_i Y. \tag{36}$$

It is therefore in fact sufficient to determine all the possibilities for some ratio β_j/β_i with $\alpha_i \neq \alpha_j$, and investigate when these lead to solutions X, Y in $\mathcal{O}$. We shall consider separately various types of Thue equations, depending on the parameters $\alpha_1,\ldots,\alpha_n,\mu$. As in the case of characteristic zero in Chapter II, §1, it will be convenient to denote by λ_i the cross-ratio $(\alpha_3-\alpha_i)(\alpha_1-\alpha_2)/(\alpha_i-\alpha_2)(\alpha_3-\alpha_1)$ for $1\le i\le n$, where we write $\lambda_i=\infty$ if $\alpha_i=\alpha_2$.

The first type of Thue equation we shall consider occurs when at least one of the finite cross-ratios λ_i is not an element of k; here we shall establish that (9) has only finitely many solutions in $\mathcal{O}$. For then the extension $K/k(\lambda_i)$ is finite, of degree $H(\lambda_i)$, and so has finite inseparability degree. From the equivalence of the conditions (34) in Chapter VI we deduce that, for some power of p, q say, λ_i is not a q-th power in K. However, the quotient of $\gamma_1=\beta_2(\alpha_3-\alpha_i)/\beta_3(\alpha_2-\alpha_i)$ and $\gamma_2=\beta_2(\alpha_3-\alpha_1)/\beta_3(\alpha_2-\alpha_1)$ is λ_i, so we deduce that at least one of γ_1 and γ_2 is not a q-th power in K, say the latter. From Lemma 11, as applied above, we conclude that γ_2 has only finitely many possibilities in K which are effectively determinable since γ_2 neither lies in k, nor is equal to any $\eta_r^{q'}$ for $q'\ge q$. Hence X and Y have only finitely many possibilities in $\mathcal{O}$, and these are effectively determinable, as required. Theorem 13 is thus established in this case.

Henceforth we shall assume, as we may, that all the finite cross-ratios λ_i lie in k. In this case we transform the Thue equation (9) by the substitution

$$X = \alpha_3\rho x + \alpha_2\tau y, \qquad Y = \rho x + \tau y, \tag{12}$$

where

$$\rho = \pi\frac{\alpha_1-\alpha_2}{\alpha_3-\alpha_2}, \qquad \tau = \pi\frac{\alpha_3-\alpha_1}{\alpha_3-\alpha_2},$$

and π is some non-zero element of K to be specified later. If α_i differs from both α_1 and α_2, then $X-\alpha_iY=\pi(\alpha_1-\alpha_i)(u_ix+v_iy)$, where $u_i=\lambda_i/(\lambda_i-1)$, $v_i=1/(1-\lambda_i)$. Furthermore, $X-\alpha_2Y=\pi(\alpha_1-\alpha_2)x$, so we choose $u_i=1$, $v_i=0$ if $\alpha_i=\alpha_2$; $X-\alpha_1Y=\pi(\alpha_1-\alpha_2)(\alpha_3-\alpha_1)(\alpha_3-\alpha_2)^{-1}(x-y)$, so we choose $u_i=-v_i=1$ if $\alpha_i=\alpha_1$. The substitution (12) thus transforms the Thue equation (9) into

$$\pi^n\prod_{i=1}^{n}(u_ix+v_iy) = \phi, \tag{37}$$

where ϕ is some fixed element of K and u_i, v_i lie in k for $1 \le i \le n$. Furthermore, we have $\gamma_2 = \beta_2(\alpha_3-\alpha_1)/\beta_3(\alpha_2-\alpha_1) = x/y$, so the hypotheses of Lemma 11 apply with $\alpha = x/y$, $\nu = \mu(\alpha_2-\alpha_3)(\alpha_3-\alpha_1)(\alpha_1-\alpha_2)$ and we deduce that x/y is either an element of k, or equal to some unit η_i^q, $1 \le i \le s$, $q = p^j$, $j \ge 0$, or a member of some finite effectively determinable subset of K. Furthermore, the hypotheses of Lemma 11 also apply with $\alpha = \lambda_i x/y$, $\nu = \mu(\alpha_2-\alpha_3)(\alpha_3-\alpha_i)(\alpha_i-\alpha_2)$, $\alpha_i \neq \alpha_2$ or α_3, and so the same trichotomy obtains for each $\lambda_i x/y$. We shall treat separately each of the three different ranges of possible values for x/y; it is the second in which the analysis differs from that of characteristic zero.

The last range of possibilities for x/y may be dealt with instantly; here we choose $\pi = 1$. For if x/y may be determined effectively, then so may x and y from (37), up to a factor of an n-th root of unity. The substitution (12) then yields X and Y, and if these lie in $\mathcal{O}$ they give a solution in $\mathcal{O}$ of (9). We conclude that if x/y is a member of some finite effectively determinable subset of K, then X and Y have only finitely many possibilities which are effectively determinable. The same conclusion applies when some $\lambda_i x/y$ is a member of a finite effectively determinable subset of K, for λ_i is a fixed constant and so x/y has only finitely many possibilities. The third range of possibilities has thus been disposed of.

We now deal with the first range of possibilities, when x/y lies in the ground field k. Since u_i and v_i also lie in k for $1 \le i \le n$, so does $y^{-n}\prod_{i=1}^{n}(u_i x + v_i y)$, that is, $\phi/(\pi y)^n$. Hence if any solutions exist with x/y in k, then ϕ is an n-th power in K; when this occurs we may choose π with $\pi^n = \phi$, and then from (37) x and y are both elements of k. We now claim that unless ρ and τ in (12) are elements of $\mathcal{O}$ then x/y may be effectively determined. For if there exists a finite valuation v on K with $v(\rho) < 0$ or $v(\tau) < 0$, then just as in the case of characteristic zero in Chapter II we consider the Puiseux expansions of ρ and τ at v, namely

$$\sigma_v(\rho) = \sum_{h=v(\rho)}^{\infty} a_h z_v^h, \qquad \sigma_v(\tau) = \sum_{h=v(\tau)}^{\infty} b_h z_v^h,$$

where a_h, b_h are elements of k. Since $Y = \rho x + \tau y$ is an element of $\mathcal{O}$, and x, y lie in k, we have $v(Y) \ge 0$ and so $a_h x + b_h y = 0$ for $h < 0$. This equation is non-trivial if $h = \min(v(\rho), v(\tau))$ and so the ratio x/y is determined as required. As above, X and Y are then determined up to a factor of an n-th root of unity. If, on the other hand, ρ and τ are both elements of $\mathcal{O}$, then x and y may assume any values in k such that $\prod_{i=1}^{n}(u_i x+v_i y)=1$; X and Y, given by (12), will then form a solution in $\mathcal{O}$ of the Thue equation (9). Since k is algebraically closed, there are infinitely many possibilities for x and y, so this yields an infinite family of solutions in $\mathcal{O}$ of (9).

Now each λ_i lies in k, so if some $\lambda_i x/y$, for $\alpha_i \neq \alpha_2, \alpha_3$, is an element of k, then so is x/y. We conclude that the only remaining case to be considered is when each $\lambda_i x/y$ for $\alpha_i \neq \alpha_2, \alpha_3$ is equal to some unit $\eta_{j_i}^{q_i}$, $1 \le j_i \le s$, $q_i = p^{t_i}$, $t_i \ge 0$. We shall assume throughout the remaining part of this section that this assumption holds. Since there are only a finite number of possibilities for each of the integers j_i, we shall further assume that these are fixed; our object is to determine the range of possible values for the integers t_i. Now if $x/y = \eta_j^q$ for some power q of p, then since η_j is not a p-th power in K, the inseparability degree of the extension K/k(x/y) is equal to q. However, λ_i lies in k, so the inseparability degree of $K/k(\lambda_i x/y)$ is also equal to q, and thus $\lambda_i x/y = \eta_{j_i}^q$: that is, all the exponents t_i are equal. Hence $\lambda_i x/y = \eta_{j_i}^q$ for some $q = p^t$, $t \ge 0$ and each $\alpha_i \neq \alpha_2, \alpha_3$; we wish to determine the range of possible values for t.

Let us now consider the case when, although all the finite cross-ratios λ_i lie in k, not all have finite multiplicative order, that is, some λ_i is transcendental over the finite field $\mathbb{F}_p$. In this case q is uniquely determined, for we have $\lambda_i = (\eta_{j_i}/\eta_{j_1})^q$, an equation in q with at most one solution. Furthermore, we may determine effectively whether such a solution exists and compute its value when it does, by virtue of the assumption made in Chapter VI on the form of presentation of the ground field k. Theorem 13 has now been established in this case also.

The remaining case to consider is when each λ_i has finite multiplicative order. We shall show that provided t is sufficiently large, it belongs to a certain congruence class of integers, which may be determined effectively: this will complete the proof of Theorem 13. We shall treat separately the three conditions, expressed by the equations $\lambda_i x/y = \eta_{j_i}^{p^t}$, the equation (37), and that X, Y are integral over k[z]; the restriction each places on t will be determined. Let m denote the smallest positive integer such that $\lambda_i^{p^m} = \lambda_i$ for each finite λ_i; such an m may be computed, and we claim that any possible values of t lie in a certain congruence class modulo m. For $\lambda_i = (\eta_{j_i}/\eta_{j_1})^{p^t}$ whenever α_i differs from α_1, α_2 and α_3, so η_{j_i}/η_{j_1} must also have finite multiplicative order. Furthermore, it may be determined by the assumption on the form of presentation of k whether these equations have a solution $t = t_o$; if so, all other solutions are given by $t \equiv t_o \pmod m$, as required.

Let us now consider the restriction on t equivalent to the equation (37). Since $u_i = \lambda_i/(\lambda_i - 1)$ and $v_i = 1/(1-\lambda_i)$ for $\alpha_i \neq \alpha_1, \alpha_2$, we have $u_i x + v_i y = (1-\eta_{j_i})^q y/(1-\lambda_i)$; if $\alpha_i = \alpha_2$ we have $u_i x + v_i y = x = \eta_{j_1}^q y$, and if $\alpha_i = \alpha_1$ we have $u_i x + v_i y = x - y = (\eta_{j_1} - 1)^q y$. Now each $1-\eta_{j_i}$ is also a unit in $\mathcal{O}$, so (37) is transformed into the equation $\pi^n \lambda y^n \eta^q = \phi$, where λ lies in k and η is some fixed unit in $\mathcal{O}$. Thus (37) has a solution y in K provided η^q/ϕ is an n-th power in K. Let us write $n_o = [K(\eta^{1/n}):K]$, $n_1 = n/n_o$, and $n_o = n_2 p^{n_3}$, where $(n_2, p) = 1$. We shall show that the condition that η^q/ϕ is an n-th power in K has either no solutions for $q = p^t$, or one solution with $t < n_3$, or a certain congruence class of solutions with $t \geq n_3$. We have $\eta = \varepsilon^{n_1}$ for some ε in K, so ϕ is necessarily an n_1-th power in K if any solutions exist. Now let us suppose that two solutions $q = p^t$, $q' = p^{t'}$ exist, with $t < t'$ say. It follows that $\eta^{q-q'}$ is an n-th power in K, so n_o divides $q-q'$, and hence $t \geq n_3$ and $p^{t'-t} \equiv 1 \pmod{n_2}$. We conclude that if η^q/ϕ is an n-th power in K for some $q = p^t$, $t < n_3$, then this is the only possible value of q. Furthermore, we conclude that if η^q/ϕ is an n-th power in K for some $q = p^{t_o}$, $n_3 \leq t_o < n_3 + n_4$, where n_4 is the multiplicative order of $p \pmod{n_2}$, then such a t_o is unique and all solutions $q = p^t$ are given $t \equiv t_o \pmod{n_4}$, $t \geq t_o$. Finally, if η^q/ϕ is not an n-th power in K for any $q = p^t$, $0 \leq t < n_3 + n_4$, then no solutions exist. It may of course be determined effectively whether each η^q/ϕ is an n-th power in K, by virtue of Lemma 1. Hence the restriction on t equivalent to (37) may be effectively determined, and there is either no solution, one solution, or a congruence class of solutions sufficiently large.

Now if t satisfies the conditions already discovered, then $(\eta_{j_i}/\eta_{j_1})^q = \lambda_i$ for each $\alpha_i \neq \alpha_1, \alpha_2, \alpha_3$, and η^q/ϕ is an n-th power in K. Since λ lies in k, the equation $y^n = \phi/\eta^q\lambda$ is soluble for y, and $x/y = \eta_{j_1}^q$; we choose $\pi = 1$ here. Hence it only remains to be determined for which values of t the solution X, Y in K of (9), given by the substitution (12), actually lies in $\mathcal{O}$. We shall show that, provided t is sufficiently large, this final condition is equivalent to a congruence equation for t. This, together with the restrictions above, provides the complete range of admissible values for t and so completes the proof of Theorem 13. Let us first rewrite (12) in the form

$$Y = y\frac{\alpha_1 - \alpha_2}{\alpha_3 - \alpha_2}\left(\frac{x}{y} + \frac{\alpha_3 - \alpha_1}{\alpha_1 - \alpha_2}\right);$$

we observe that if Y lies in $\mathcal{O}$, then, by the Thue equation (9), so does X. We need therefore only investigate the restrictions on t equivalent to the restrictions $v(Y) \geq 0$ for each finite valuation v on K. Let us denote for brevity by ψ and ξ the elements $\phi((\alpha_1-\alpha_2)/(\alpha_3-\alpha_2))^n$ and $(\alpha_3-\alpha_1)/(\alpha_1-\alpha_2)$ respectively. In view of the equation $y^n = \phi/\eta^q\lambda$, where λ is an element of k and η is some unit of $\mathcal{O}$, it suffices to determine for which values of $t \geq 0$ the inequalities

$$v(\theta^{p^t} - \xi) \geq -v(\psi)/n \tag{38}$$

hold for each $v \nmid \infty$; here we write $\theta = \eta_{j_1}$ for brevity. Since θ is a unit of $\mathcal{O}$ we have $v(\theta) = 0$ whenever $v \nmid \infty$, so if v also satisfies $v(\xi) > 0$, then we require $v(\psi) \geq 0$, whilst if $v(\xi) < 0$ we require $v(\psi) + nv(\xi) \geq 0$; both conditions are independent of t. It remains to investigate those valuations $v \nmid \infty$ for which $v(\xi) = 0$ and $v(\psi) < 0$; these are only finite in number. We shall show that, provided $np^t \geq -v(\psi)$, (38) has either no solutions for t, one solution, or a certain congruence class of solutions. This demonstration will complete the proof of Theorem 13, since there are only finitely many such inequalities (38) which need to be considered. For each such valuation v, let us denote by $\theta(v)$ and $\xi(v)$ the constant coefficient in the Puiseux expansion at v of θ and ξ respectively, so $v(\theta - \theta(v)) > 0$, $v(\xi - \xi(v)) > 0$. Since k has characteristic p, we have $v(\theta^{p^t} - \theta(v)^{p^t}) = p^t v(\theta - \theta(v)) \geq p^t$, so provided $np^t \geq -v(\psi)$ the inequality (38) is equivalent to

$$v(\theta(v)^{p^t} - \xi) \geq -v(\psi)/n. \tag{39}$$

If $v(\xi - \xi(v)) < -v(\psi)/n$ then (39) has no solutions for t, whilst if $v(\xi - \xi(v)) \geq -v(\psi)/n$ then (39) is equivalent to the equation

$$\theta(v)^{p^t} = \xi(v).$$

It may be determined effectively, by virtue of the assumption on the form of presentation of k, whether this final equation has any solutions. Furthermore, if it has, then there is either one solution or a single congruence class of solutions for t, according as the multiplicative order of $\theta(v)$ in k is infinite or finite. We have now established that the range of admissible values for t is either empty, or finite, or it consists of a finite set of possibilities for $t < M$, together with a single congruence class with $t \geq M$. Furthermore, the congruence class, the integer M, and the finite number of exceptional values, are all effectively determinable.

The proof of Theorem 13 is now complete, and in the next section we shall illustrate by a specific example the general algorithm proved here. However, it may perhaps be of some interest to examine further the significance of the range of values of t determined above. If we choose π in the substitution (12) such that $\pi^n = \phi$, then although π lies in an extension of K, the Thue equation (9) has now been transformed into $F(x,y) = 1$, where $F(x,y) = \prod_{i=1}^{n} (u_i x + v_i y)$ is a form with coefficients in the ground field k. Furthermore, since $\lambda_i^{p^m} = \lambda_i$ for each finite λ_i, we have $F(x,y)^{q_0} = F(x^{q_0}, y^{q_0})$, where $q_0 = p^m$. Thus whenever x, y is a solution in K of $F(x,y) = 1$, so is x^{q_0}, y^{q_0}. We may now summarise the results of this section as they appertain to Theorem 13 as follows, in which all the conditions and specific solutions are effectively determinable; we recall that λ_i denotes the cross-ratio $(\alpha_3-\alpha_i)(\alpha_1-\alpha_2)/(\alpha_i-\alpha_2)(\alpha_3-\alpha_1)$, and we will be concerned with λ_i only when α_i differs from α_1, α_2 and α_3. If some λ_i is not an element of k then the Thue equation (9) has only finitely many solutions X, Y in $\mathcal{O}$. If each λ_i is an element of k then we transform (9) by the substitution (12), with $\pi^n = \phi$. If the coefficients of the substitution (12) lie in $\mathcal{O}$, then x and y may assume any values in k such that $F(x,y) = 1$; if not, then only finitely many solutions x, y in k of $F(x,y) = 1$ lead to solutions X, Y in $\mathcal{O}$ of (9). This deals with x, y in k; let us now exclude the solutions X, Y corresponding to these values. If some λ_i has infinite multiplicative order in k then X and Y have only finitely many possibilities in $\mathcal{O}$. If each λ_i has finite multiplicative order then there exists a power q_0 of p such that $\lambda_i^{q_0} = \lambda_i$ for each λ_i. Furthermore, there now exists a (possibly empty) finite set $x_1, y_1; \ldots; x_N, y_N$ of solutions of $F(x,y) = 1$, together with powers $q_1, \ldots, q_N$ of q_0 such that choosing $x = x_i^{q_i^t}$, $y = y_i^{q_i^t}$ in (12) for any

$1 \le i \le N$ and any $t \ge 0$ leads to a solution X, Y in $\mathcal{O}$ of (9). Furthermore, apart from these infinite families there are at most finitely many solutions in $\mathcal{O}$ of the Thue equation (9).

2 AN EXAMPLE

In this section we shall determine all the solutions X, Y of the equation

$$X(X-Y)(X+Y) = 1$$

such that X, Y are integral over $k[z]$ and lie in the field K, generated over $k(z)$ by y and t satisfying

$$t^2 + tz + 1 = 0, \quad y^2 = t^3 - 1,$$

where k is an algebraically closed field of characteristic 3. This will illustrate the general method of solution described in §1. We observe that, as for the example in Chapter II, §3 in the case of characteristic zero, no machine computation is necessary: this is a consequence of the sharpness of the fundamental inequality (11). We observe that whenever X, Y is a solution in $\mathcal{O}$ of $X(X-Y)(X+Y)=1$, so also is X^3, Y^3. In accordance with the discovery of the various families of solutions in the analysis in §1, we shall determine a set $x_1, y_1; \ldots; x_M, y_M$ of solutions in $\mathcal{O}$ such that any solution X, Y in $\mathcal{O}$ of $X(X-Y)(X+Y)=1$ either lies in k, or is given as $X = x_i^{3^T}$, $Y = y_i^{3^T}$ for some $1 \le i \le M$ and $T \ge 0$.

The equations $t^2 + tz + 1 = 0$ and $y^2 = t^3 - 1$ are separable for t and y respectively, so $K/k(z)$ is separable and z is not a cube in K. The genus may be calculated in a similar fashion to that in Chapter II, §3. Alternatively, if we observe that $y^2 = (t-1)^3$, then we see that $K = k(y/(t-1))$, and so K/k is rational and thus has genus zero. The expansions of the element $y/(t-1)$ which correspond to the infinite valuations on K are

$$\chi(z_1^{-1} - z_1 - \ldots), \quad \chi(1 - z_2 - \ldots), \quad -\chi(1 - z_3 - \ldots),$$

respectively, where χ is an element of k satisfying $\chi^2 = -1$, and the local parameter z_i is equal to $z^{-1/2}$ for i=1 and z^{-1} for i=2,3. Hence there are three infinite valuations on K with ramification indices 2,1,1 respectively. As before we write $\beta_1 = X$, $\beta_2 = X-Y$, $\beta_3 = X+Y$, and thereby obtain the equations

$$\beta_1\beta_2\beta_3 = 1 \quad , \quad \beta_1 + \beta_2 + \beta_3 = 0.$$

Furthermore, β_1, β_2, β_3 are elements of $\mathcal{O}$ and so by the first relation they are in fact units of $\mathcal{O}$, so $v(\beta_i) = 0$ for each $v \nmid \infty$. If $\alpha = -\beta_1/\beta_3$, then α satisfies the hypotheses of Lemma 11 with $\nu = 1$. We conclude that α is either an element of k, or it is equal to $\eta_i^{3^T}$ for some i,T, $1 \le i \le s, T \ge 0$, where $\eta_1,\ldots,\eta_s$ are the units in $\mathcal{O}$ which are not cubes and are such that each $1-\eta_i$ is also a unit. If α is an element of k then we obtain the infinite family of solutions X, Y in k. If $-\beta_1/\beta_3 = \eta_i^{3^T}$, then $\beta_2/\beta_3 = -(1-\eta_i)^{3^T}$, and so $\beta_3^3(\eta_i(1-\eta_i))^{3^T} = 1$. Now $\eta_i(1-\eta_i)$ is not a cube in K, since otherwise it would have zero derivative, in which case so also would η_i, which does not occur. Hence we conclude that $T \ge 1$, and so the complete set of non-constant solutions in $\mathcal{O}$ of $X(X-Y)(X+Y) = 1$ is given by $X = (\eta_i^2/(\eta_i-1))^q$, $Y = (1+\eta_i)^{3q}/(\eta_i(1-\eta_i))^q$ for $1 \le i \le s$, $q = 3^{T-1}$, $T \ge 1$.

It remains then, to determine $\eta_1,\ldots,\eta_s$, that is, the set of units η of $\mathcal{O}$ such that $1-\eta$ is also a unit, but η is not a cube. In view of the fundamental inequality (11) we have $H(\eta) \le 1$, and since η is not an element of k we have $H(\eta) = 1$. Since $H(1-\eta) = 1$ also, we deduce that there are only a finite number of possibilities for the integers $v_i(\eta)$, $v_i(1-\eta)$, $i = 1,2,3$. Taking into account the sum formula (1) which yields $\sum_{i=1}^{3} v_i(\eta) = \sum_{i=1}^{3} v_i(1-\eta) = 0$, we conclude that the only possibilities are

$$v_i(\eta) = v_i(1-\eta) = -1 \quad , \quad v_j(\eta) = v_\ell(1-\eta) = 0 \quad ,$$

$$v_\ell(\eta) = v_j(1-\eta) = 1$$

for some permutation i,j,ℓ of 1,2,3. However, whenever η is a unit satisfying the conditions that $1-\eta$ is a unit and η is not a cube, the conditions are also satisfied by $1-\eta$, $1/\eta$, $1-1/\eta$, $1/(1-\eta)$ and $\eta/(\eta-1)$. Hence we may in fact assume that $i=1$, $j=2$, $\ell=3$; the five other permutations of 1,2,3 then correspond to the five other units.

We wish to determine α and β such that

$$v_1(\alpha) = v_1(\beta) = -1 \ , \quad v_2(\alpha) = v_3(\beta) = 0 \ , \quad v_3(\alpha) = v_2(\beta) = 1 \ ,$$

and $v(\alpha) = v(\beta) = 0$ if $v \nmid \infty$.

If α and β are so determined, then η/α and $(1-\eta)/\beta$ have value zero at every valuation on K, and so are elements of k, equal to a and b respectively, say. Since $a\alpha + b\beta = 1$ we obtain $a\alpha' + b\beta' = 0$, and thus a and b are determined, as $\alpha' \neq 0 \neq \beta'$. The computation of elements α and β may be achieved using Lemma 1, just as in Chapter II, §. However, the argument may be considerably abbreviated by the observation that K is just the rational function field generated over k by $Z = y/(t-1)$. Hence any element f of K may be written in the form $P(Z)/Q(Z)$, where P and Q are coprime polynomials in $k[z]$. We now have $H(f) = \max(\deg P, \deg Q)$, and the valuations v_1, v_2, v_3 on K are respectively equal to the valuations ord, ord_χ, $\mathrm{ord}_{-\chi}$ on the rational function field $k(Z)$. The conditions above on α and β give $\alpha = Z + \chi$, $\beta = Z - \chi$, up to constant factors. Thus $a+b = 0$ and $(a-b)\chi = 1$, so $a = -b = \chi$, and $\eta = \chi Z - 1$. The other five units $1-\eta$, $1/\eta$, $1-1/\eta$, $1/(1-\eta)$ and $\eta/(\eta-1)$ may be calculated, and we deduce that the non-constant solutions in $\mathcal{O}$ of $X(X-Y)(X+Y) = 1$ are

$$Y = \pm(\chi y/t)^q \ , \quad X = t^{-q} \quad \text{or} \quad ((1 \pm \chi y)/t)^q$$

where χ is an element of k satisfying $\chi^2 = -1$ and $q = 3^j$ for any $j \geq 0$.

If k has characteristic $p > 3$, then if X and Y are not p-th powers in K the analysis to determine solutions in $\mathcal{O}$ of $X(X-Y)(X+Y) = 1$ proceeds just as for characteristic zero in Chapter II, §3. We deduce that the non-constant solutions in $\mathcal{O}$ are

$$X = (\omega/t)^q, \quad Y = (\omega\chi y/t)^q \text{ and } X = (\omega(\chi y-1)/2t)^q,$$

$$Y = \pm(\omega(\chi y+3)/2t)^q,$$

where χ and ω take all values in k such that $\chi^2 = -1$, $\omega^3 = 1$, and $q = p^j$ for any $j \geq 0$.

For completeness we include the case of characteristic 2, although, since the factors X, X-Y, X+Y are no longer distinct, $X(X-Y)(X+Y) = 1$ is not a Thue equation. The solutions in $\mathcal{O}$ are evidently given by

$$X = \eta^{-2} \quad , \quad Y = \eta + \eta^{-2}$$

for any unit of $\mathcal{O}$. It thus suffices to determine the group of units of $\mathcal{O}$. Although it is not possible to do this effectively for an extension of $k(z)$ of arbitrary genus, in this case K/k has genus zero since $K = k(x)$ where $x = (y+1)/t$. We have $z = (x+1/x)^2$, so the 2 infinite valuations on K are just the valuations ord and ord_0 on the rational function field $k(x)$. It follows that any unit η in $\mathcal{O}$ may be written in the form $\eta = ax^n$ where a is any non-zero element of k and n is any integer. This provides the complete set of solutions in $\mathcal{O}$ to $X(X-Y)(X+Y) = 1$ when k has characteristic 2.

3 *THE HYPERELLIPTIC EQUATION*

In this section we shall solve completely the general hyperelliptic equation over a function field of arbitrary characteristic. As for the Thue equation in §1 it is Lemma 11 which plays the crucial role in the analysis. Let us denote by $\alpha_1,\ldots,\alpha_n$ $n(\geq 3)$ distinct elements of $\mathcal{O}$; we shall establish here the following theorem.

Theorem 14

All the solutions X, Y in $\mathcal{O}$ of the hyperelliptic equation

$$Y^2 = (X-\alpha_1)\ldots(X-\alpha_n) \tag{14}$$

may be determined effectively.

When k has characteristic 2 the equation (14) is inseparable for Y and represents a curve of genus 0 with just one infinite valuation; for any choice of X in $\mathcal{O}$, Y lies in the fixed extension field $K(\sqrt{z})$ of K. We therefore make the assumption that k has characteristic $p > 2$; the case of characteristic zero has already been solved in Chapter III. As for the Thue equation over fields of positive characteristic, it is possible for the solutions of (14) to have unbounded height, in contrast with Theorem 6 which applies to characteristic zero. For example, if $\alpha_i^p = \alpha_i$ for each i and q is any power of p, then X^q, Y^q is a solution in $\mathcal{O}$ of (14) whenever X, Y is, and $H(X^q) = qH(X)$, so the heights are unbounded if a non-constant integral solution exists. We shall demonstrate that (14) may have several infinite families of solutions of this type, together with an infinite family of solutions derived from the solutions in k of a hyperelliptic equation over k as in Theorem 5, together with a finite number of exceptional solutions. The strategy of proof is to combine the approach used in Chapter III for characteristic zero with Lemma 11.

Let X, Y denote a solution in $\mathcal{O}$ of (14), and then let us denote by L the extension field of K obtained by adjoining the square roots of $X-\alpha_1$, $X-\alpha_2$ and $X-\alpha_3$. As in §3 of Chapter III, we wish to show there are only finitely many effectively determinable possibilities for L as X, Y run over the solutions in $\mathcal{O}$ of (14). We note first that the proof that L has only finitely many possibilities carries over to the case of characteristic $p \neq 2$ without alteration (see [30] and [18]). For the effective determination of these possibilities for L, we shall apply the same method as that in §3 of Chapter III, which applies with just one alteration, which we now discuss. At the end of the proof we desired to express the coefficients in the expansion of an element f of K at a finite unramified valuation v in terms of the constant coefficients of the derivatives of f. When k has characteristic zero there is no difficulty, for if σ_v denotes the embedding corresponding to the valuation $v|a$ in the field $k\{z-a\}$, then $\sigma_v(f) = \sum_{i=0}^{\infty} c_i(z-a)^i$ for elements c_i, $i \geq 0$, in k, and so $(i!)c_i = f^{(i)}(v)$, where $f^{(i)}$ denotes the i-th derivative of f with respect to z, and $f^{(i)}(v)$ is the constant coefficient of $f^{(i)}$ at v. Hence $c_i = 0$ if and only if $f^{(i)}(v) = 0$ when k has characteristic zero. The analysis is not quite as straightforward if k has positive characteristic p, since $i! = 0$ for $i \geq p$; nevertheless we recover the coefficients c_i for $0 \leq i < p$ in this way. Our object is to recover the higher coefficients of a fixed element f in K in a fashion independent of the unknown valuation v. If K^p denotes the field of p-th powers in K, then the extension K/K^p has degree p, and since z is not in K^p we may write

$$f = f_1^p + f_2^p z + \ldots + f_p^p z^{p-1} \tag{40}$$

for some unique elements $f_1,\ldots,f_p$ in K. In fact $f_1,\ldots,f_p$ may be determined by differentiating (40) up to p-1 times; we obtain $f_p^p = -f^{(p-1)}$, $f_{p-1}^p = f^{(p-2)} - f^{(p-1)}z$ and so on. We thereby obtain $\sigma_v(f_p^p) = \sum_{i=0}^{\infty} c_{ip+p-1}(z-a)^{ip}$, and so, upon taking p-th roots,

$\sigma_v(f_p) = \sum_{i=0}^{\infty} c_{ip+p-1}^{1/p}(z-a)^i$. Differentiating this equation up to p-1 times as above, we realise that $i!c_{ip+p-1} = (f_p^{(i)}(v))^p$, so if $0 \le i < p$, then $c_{ip+p-1} = 0$ if and only if $f_p^{(i)}(v) = 0$. Similarly if $0 \le i < p$ then $c_{ip+p-2} = 0$ if and only if $h(v) = 0$, where $h = (f_p^{(i)})^p - (f_{p-1}^{(i)})^p z$. From the expressions for $f_{p-2}, \ldots, f_1$ we may recover all the coefficients c_i with $0 \le i < p^2$. By repeating this process, expressing each of $f_1, \ldots, f_p$ in the form (40), we may recover the coefficients c_i for $0 \le i < p^3$ and so eventually all the coefficients c_i as required. The remainder of the proof that the possibilities for L may be determined effectively follows just as in the case of characteristic zero in §3 of Chapter III. We shall assume henceforth that $X-\alpha_1$, $X-\alpha_2$, $X-\alpha_3$ are squares in a fixed field L. Under this assumption we shall proceed to determine effectively the solutions X, Y in $\mathcal{O}$ of the hyperelliptic equation (14).

As in Chapter III we define elements $\beta_1 = \xi_2 - \xi_3$, $\hat{\beta}_1 = \xi_2 + \xi_3$, with $\beta_2, \hat{\beta}_2, \beta_3$ and $\hat{\beta}_3$ defined similarly by permutation of the suffixes; ξ_1, ξ_2 and ξ_3 denote the square roots in L of $X-\alpha_1$, $X-\alpha_2$ and $X-\alpha_3$ respectively. Then β_1, $\hat{\beta}_1$, β_2, $\hat{\beta}_2$, β_3 and $\hat{\beta}_3$ are all elements of $\mathcal{O}_L$, the ring of elements of L integral over $k[z]$. Furthermore, $\beta_1\hat{\beta}_1 = \alpha_3 - \alpha_2$, $\beta_2\hat{\beta}_2 = \alpha_1 - \alpha_3$, $\beta_3\hat{\beta}_3 = \alpha_2 - \alpha_1$,

$$\beta_1 + \beta_2 + \beta_3 = 0 \tag{15}$$

and

$$\beta_1 + \hat{\beta}_2 - \hat{\beta}_3 = -\hat{\beta}_1 + \beta_2 + \hat{\beta}_3 = \hat{\beta}_1 - \hat{\beta}_2 + \beta_3 = 0. \tag{16}$$

Lemma 11 may now be applied directly to L, with $\alpha = -\beta_1/\beta_2$ and $\nu = (\alpha_3-\alpha_2)(\alpha_1-\alpha_3)(\alpha_2-\alpha_3)$. If we denote by $\eta_1, \ldots, \eta_s$ the finite computable subset of L such that no η_i is a p-th power in L and both η_i and $1-\eta_i$ are units in $\mathcal{O}_L$ for $1 \le i \le s$, then we conclude from Lemma 11 that $-\beta_1/\beta_2$ is either an element of k, or it is equal to some $\eta_i^{p^j}$, $1 \le i \le s$, $j \ge 0$, or it is a member of some finite computable subset of L. As in Chapter III we note that $\beta_1/\beta_2 = f$ determines X uniquely, for if we write $\kappa = (\alpha_1-\alpha_3)/(\alpha_2-\alpha_1)$ for brevity, then $\beta_3/\hat{\beta}_3 = -(1+f)(1+\kappa+\kappa f)/f$, and

$$(X-\alpha_1)/(\alpha_2-\alpha_1) = \xi_1^2/\beta_3\hat{\beta}_3 = \tfrac{1}{4}(\beta_3+\hat{\beta}_3)^2/\beta_3\hat{\beta}_3, \tag{17}$$

so f determines X as required. As usual we shall treat separately several types of hyperelliptic equation, depending on the parameters $\alpha_1,\ldots,\alpha_n$. For convenience we shall denote the ratio $(\alpha_1-\alpha_i)/(\alpha_2-\alpha_1)$ by κ_i, $1\le i\le n$.

The first case we consider occurs when some κ_i is not an element of the ground field k. We shall prove that in this case the hyperelliptic equation (14) has but finitely many solutions in $\mathcal{O}$ which may be determined effectively. Rearranging $\alpha_3,\ldots,\alpha_n$ if necessary, we may assume that in fact $\kappa_3=\kappa$ is not an element of k. Hence the extension $K/k(\kappa)$ is finite, of degree $H(\kappa)$, and so has finite inseparability degree. We deduce from the equivalence of the conditions (34) that, for some power q of p, κ is not a q-th power in K; since L/K is separable κ is neither a q-th power in L. However, the product of β_1/β_2 and $\hat{\beta}_1/\hat{\beta}_2$ is $-1-1/\kappa$, so at least one of β_1/β_2 and $\hat{\beta}_1/\hat{\beta}_2$ is not a q-th power in L. In view of Lemma 11, applied with α equal to either $-\beta_1/\beta_2$ or $\hat{\beta}_1/\hat{\beta}_2$ as appropriate and $\nu=(\alpha_3-\alpha_2)(\alpha_1-\alpha_3)(\alpha_2-\alpha_1)$, we conclude that β_1/β_2 has only a finite number of possibilities, and these may be determined effectively. Since we observed above that X is determined uniquely by β_1/β_2, we conclude that Theorem 14 is established in this case, and (14) has only finitely many solutions in $\mathcal{O}$.

Henceforth we shall assume that each κ_i, $1\le i\le n$, lies in the ground field k. In this case we make the substitution $X=(\alpha_2-\alpha_1)x+\alpha_1$, which transforms (14) into

$$Y^2 = (\alpha_2-\alpha_1)^n\prod_{i=1}^{n}(x+\kappa_i). \tag{41}$$

Furthermore, if $\alpha=-\beta_1/\beta_2$ or $\hat{\beta}_1/\hat{\beta}_2$, then α is either an element of k, or it is equal to some $\eta_i^{p^t}$, $1\le i\le s$, $t\ge 0$, or it is a member of some finite computable subset of L. The final part of the trichotomy may be dealt with as above, since the product of β_1/β_2 and $\hat{\beta}_1/\hat{\beta}_2$ is fixed, and so either quantity determines X uniquely. This final possibility therefore contributes only finitely many solutions of (14), all of which may be computed. Let us now examine the first part, when either β_1/β_2 or $\hat{\beta}_1/\hat{\beta}_2$, and hence both, lie in k. In this case we conclude that $\beta_3/\hat{\beta}_3$ also lies in k and so even also does x in view of (17). Unless Y vanishes, (41) possesses a solution x in k, Y in $\mathcal{O}$ if and only if $(\alpha_2-\alpha_1)^n$ is a square in $\mathcal{O}$. If this occurs, say $(\alpha_2-\alpha_1)^n=\gamma^2$, then the substitution $Y=\gamma y$

further transforms (41) into $y^2 = \prod_{i=1}^{n}(x+\kappa_i)$, an equation with coefficients in the ground field k. The solutions X, Y in $\mathcal{O}$ of (14) with β_1/β_2 in k are thus placed in correspondence with the infinite family of solutions in k of $y^2 = \prod_{i=1}^{n}(x+\kappa_i)$ via the substitution $X = (\alpha_2-\alpha_1)x+\alpha_1$, $Y = \gamma y$ with coefficients in $\mathcal{O}$.

Henceforth we shall assume that the central part of the trichotomy obtains, so that $-\beta_1/\beta_2 = \eta_i^q$ and $\hat{\beta}_1/\hat{\beta}_2 = \eta_j^{q'}$ for some i,j, $1 \le i,j \le s$, and powers q,q' of p. Now i and j have only finitely many possibilities, so we shall suppose henceforth that these are fixed, and it only remains to determine the range of possibilities for q and q'. Since the product $\eta_i^q\eta_j^{q'}$ is equal to $1+1/\kappa$, an element of k, we deduce that η_i^q is a q'-th power in L, and so $q' \le q$; similarly $q \le q'$ and so $q = q'$. Hence $(\eta_i\eta_j)^q = 1+1/\kappa$, and so if κ has infinite multiplicative order in k, then so also does $1+1/\kappa$, and thus q has at most one possibility, which is indeed effectively determinable by the assumption on the form of presentation of the ground field k made in Chapter VI. Furthermore, if any of $\kappa_3,\ldots,\kappa_n$ have infinite multiplicative order, then we may first rearrange $\alpha_3,\ldots,\alpha_n$ so that $\kappa_3 = \kappa$ does: hence the proof of Theorem 14 is complete in this case also.

Finally we assume that each κ_ℓ, $1 \le \ell \le n$, has finite multiplicative order in k, and that $\beta_1/\beta_2 = -\eta_i^{p^t}$, $\hat{\beta}_1/\hat{\beta}_2 = \eta_j^{p^t}$ for some i,j, $1 \le i,j \le s$; we wish to determine the range of possibilities for $t \ge 0$. First we determine a power of p, $p^m = q$ say, such that $\kappa_\ell^q = \kappa_\ell$ for $1 \le \ell \le n$. We may now write $t = mu+v$ for integers u, v, $u \ge 0$, $0 \le v < m$; since v has only finitely many possibilities we may assume that it is fixed; thus we wish to determine all possible values of $u \ge 0$ such that choosing $\beta_1/\beta_2 = \eta^{q^u}$ with $\eta = -\eta_i^{p^v}$, leads, by (15), to a solution X, Y in $\mathcal{O}$ of the hyperelliptic equation (14). We shall show that, in contrast to the rather complicated criterion in the case of the Thue equation in §1, there is the simplest dichotomy on the range of admissible values of u: either none or all are admissible. Since we may check whether $u = 0$, that is, $\beta_1/\beta_2 = \eta$, is admissible, the complete range of possibilities may be determined effectively as required. Let us denote by ε and ζ the values of x and $\prod_{i=1}^{n}(x+\kappa_i)$ respectively corresponding to $\beta_1/\beta_2 = \eta$. Since $\kappa_i^q = \kappa_i$ for each i, in view of (15) we have $x = \varepsilon^{q^u}$ and $\prod_{i=1}^{n}(x+\kappa_i) = \zeta^{q^u}$ when

$\beta_1/\beta_2 = \eta^{q^u}$. Now if X lies in K then so does ε^{q^u}, and hence so does ε as L/K is separable. Furthermore, $\zeta = \prod_{i=1}^{n} (\varepsilon+\kappa_i)$ is also an element of K and so if Y lies in K for some $u \geq 0$ then $\zeta(\alpha_2-\alpha_1)^n$ is a square in K. Conversely, we shall show that if ε lies in K and $\zeta(\alpha_2-\alpha_1)^n$ is a square in K then u may assume any non-negative value. Certainly X and Y are then elements of K, so it suffices to prove that X is always integral over $k[z]$; in view of (14) so then is Y. Indeed, since η, $1+\eta$ and $1+\kappa+\kappa\eta$ are units in $\mathcal{O}_L$, $\theta = -(1+\eta)(1+\kappa+\kappa\eta)/\eta$ is a unit in $\mathcal{O}_L$, and thus $\varepsilon = \frac{1}{4}(\theta + 2 + 1/\theta)$ is integral over $k[z]$. Hence for any $u \geq 0$ ε^{q^u} lies in $\mathcal{O}$ since ε lies in K, and so X, Y are elements of $\mathcal{O}$ as required. We have now succeeded in showing that in each case the complete range of solutions in $\mathcal{O}$ of the hyperelliptic equation (14) may be determined effectively; hence the proof of Theorem 14 is complete.

CHAPTER VIII THE SUPERELLIPTIC EQUATION

Hitherto this book has effectively resolved the problem of determining the complete set of integral solutions to certain general families of equations. The fundamental inequality has formed the crux of the argument in each analysis, and it has led to the solution in Chapter II of the Thue equation, in Chapter III of the hyperelliptic equation, and in Chapter IV of equations of genera 0 and 1. In Chapter VII we succeeded in dealing with the Thue and hyperelliptic equations over fields of positive characteristic, and it was the appropriate extension of the fundamental inequality to such fields which again provided the crucial step. In the case of positive characteristic it is possible for the heights of the integral solutions to be unbounded, but this cannot occur over fields of characteristic 0, and for that case we determined explicit bounds for each of the various families solved. The fundamental inequality contributed the essence to each of those proofs also, and it is the purpose of this concluding chapter to illustrate a further range of applications for the inequality by studying briefly the superelliptic equation. Here the inequality is employed in a rather different fashion from previously, and this new approach will in fact lead to explicit bounds on the heights of all the solutions, not just those integral. Explicit bounds for non-integral solutions have only been obtained by Schmidt *[36]* in the case of certain Thue equations, and stronger bounds may be deduced from our methods as below. A recent result *[37]* shows that it is possible, at least in principle, to determine effective bounds for solutions in rational functions of the superelliptic equation, but as with previous work this relies on an involved study of algebraic differential equations (see also *[41]*).

Let k denote an algebraically closed field of characteristic zero and K a finite extension of the rational function field $k(z)$. As discussed in the introduction, the assumption that k is algebraically closed is not necessary for the bounds established, provided in that case that K does not extend the field of constants k. We shall be concerned with the superelliptic equation

$$Y^m = f(X) , \tag{42}$$

where f is a polynomial of degree n and height H, with coefficients in K. We shall assume for convenience that f splits completely over K into distinct linear factors. It will be evident that the assumption that the factors of f are distinct could be replaced by the appropriate condition on the multiplicities at each stage, whilst combination of Lemma 4 with the theorems below would immediately yield results if f does not split over K.

Theorem 15

If X, Y is a solution in $\mathcal{O}$ of the superelliptic equation

$$Y^m = f(X) , \tag{42}$$

where $m \geq 3$ and $n \geq 2$, then

$$H(X) \leq 18H + 3(2g + r - 1) ;$$

here g and r respectively denote the genus of K/k and the number of infinite valuations on K.

Proof. Since f factorises completely in K we may write $f(X) = \alpha_o \prod_{i=1}^{n} (X-\alpha_i)$, where $\alpha_o \neq 0$ and $\alpha_1,\ldots,\alpha_n$ are distinct elements of K. The essential idea of the proof is to apply the fundamental inequality to the identity

$$(X - \alpha_1) + (\alpha_2 - X) + (\alpha_1 - \alpha_2) = 0. \tag{43}$$

The novelty of this application is that previously the set of valuations at which the three summands take different values was independent of the solution X, Y, whereas now the set depends on the particular solution, and success is gained by bounding the size of the set by a fraction of the height of X. This bound follows directly from the superelliptic equation. Clearly it suffices to take $Y \neq 0$.

We may assume that $\alpha_1,\ldots,\alpha_n$ are arranged in order of increasing height, and so

$$\sum_{i=1}^{s} H(\alpha_i) \le sH/n \qquad (s=1,\ldots,n).$$

Let us write $\beta_i = \prod_{j\neq i}(\alpha_i - \alpha_j)$, $1\le i\le n$, so that $H(\beta_i) \le H + (n-2)H(\alpha_i)$. We shall denote by $\mathcal{W}$ the set of valuations v on K at which one or more of the following occur:

$$v|\infty, \quad v(f) < 0, \quad v(\alpha_o) > 0, \quad v(\beta_1\beta_2) > 0,$$

so that

$$|\mathcal{W}| \le r + 2(3 - 2/n)H.$$

Now if v lies outside $\mathcal{W}$ then $v(\alpha_i) \ge 0$ for each i, $v(X) \ge 0$ and $v(\alpha_1-\alpha_2) = 0$. Hence if we denote by S_i, i=1,2, the set of valuations v outside $\mathcal{W}$ such that $v(X-\alpha_i) > 0$, then by the fundamental inequality (11) we obtain

$$H((X-\alpha_1)/(\alpha_1-\alpha_2)) \le |\mathcal{W}\cup S_1\cup S_2| + 2g - 1,$$

as $|\mathcal{W}| \ge r \ge 1$. Furthermore, if v lies in S_1 then since $v(X-\alpha_1) > 0$ and v lies outside $\mathcal{W}$ we have $v(X-\alpha_j) = 0$ for $2\le j\le n$ and so $v(X-\alpha_1) = mv(Y)$. Thus $v(X-\alpha_1)$ is a positive integer divisible by m, and so $m|S_1| \le H(X-\alpha_1)$. Similarly $m|S_2| \le H(X-\alpha_2)$, and so from the inequality above we obtain

$$H(X) - 2H(\alpha_1) - H(\alpha_2) \le 2H(X)/m + H(\alpha_1)/m + H(\alpha_2)/m + |\mathcal{W}| + 2g - 1.$$

Theorem 15 now follows from the bound above on $|\mathcal{W}|$, as $m \ge 3$.

It is also possible effectively to determine the solutions X, Y in $\mathcal{O}$ of $Y^m = f(X)$. For as with the hyperelliptic equation in Chapter III $X-\alpha_1$ and $X-\alpha_2$ are both m-th powers in some field L, which has only finitely many computable possibilities as X, Y range over the solutions in $\mathcal{O}$ of $Y^m = f(X)$. On writing $X-\alpha_1 = \xi_1^m$, $X-\alpha_2 = \xi_2^m$, the Thue equation $\xi_1^m - \xi_2^m = \alpha_2-\alpha_1$ may be solved as in Chapter II. We shall now turn to the more general case of solutions in K of $Y^m = f(X)$, not just in $\mathcal{O}$.

Theorem 16

If X, Y *is a solution in* K *of the superelliptic equation*

$$Y^m = f(X) \tag{42}$$

with $n=2$, $m \geq 5$ *or* $n \geq 3$, $m \geq 4$, *then*

$$H(X) \leq 35H + 8g.$$

Proof. Let us first consider the case $n=2$. Here we may apply the fundamental inequality to the identity

$$(X - \alpha_1) + (\alpha_2 - X) + (\alpha_1 - \alpha_2) = 0 \tag{43}$$

as above. If $\mathcal{W}$ now denotes the set of valuations on K at which

$$v(f) < 0, \quad v(\alpha_0) > 0 \quad \text{or} \quad v(\alpha_1 - \alpha_2) > 0,$$

then

$$|\mathcal{W}| \leq 3H.$$

Let us also denote by S_1, S_2, S_3 the sets of valuations lying outside $\mathcal{W}$ at which $v(X-\alpha_1) > 0$, $v(X-\alpha_2) > 0$, $v(X) < 0$ respectively. If v is excluded from each of the sets $\mathcal{W}$, S_1, S_2 and S_3, then $v(X-\alpha_1) = v(X-\alpha_2) = v(\alpha_1-\alpha_2) = 0$, and so the fundamental inequality (11) yields

$$H((X-\alpha_1)/(\alpha_1-\alpha_2)) \leq |\mathcal{W} \cup S_1 \cup S_2 \cup S_3| + 2g.$$

As in the proof of Theorem 15 we obtain $m|S_1| \leq H(X-\alpha_1)$ and $m|S_2| \leq H(X-\alpha_2)$. If v lies in S_3, then $v(X-\alpha_1) = v(X-\alpha_2) = v(X) < 0$, and so $mv(Y) = 2v(X)$. We conclude that if m is odd then $v(X)$ is divisible by m and so $m|S_3| \leq H(X)$, whilst if m is even then $m|S_3| \leq 2H(X)$. Hence

$$H(X)(1 - 2/m - (m,2)/m) \leq (9/2 + 1/m)H + 2g,$$

where $(m,2) = 1$ or 2 according as m is odd or even. Since $m \geq 5$, the required inequality follows. As in Theorem 15 the bound is trivial if $Y = 0$.

Let us now consider the case $n \geq 3$, so $m \geq 4$; here we shall apply the fundamental inequality to the identity

$$(\alpha_2-\alpha_3)(X-\alpha_1) + (\alpha_3-\alpha_1)(X-\alpha_2) + (\alpha_1-\alpha_2)(X-\alpha_3) = 0. \tag{44}$$

In this case we denote by $\mathcal{W}$ the set of valuations at which

$$v(f) < 0, \quad v(\alpha_o) > 0 \quad \text{or} \quad v(\beta_1\beta_2\beta_3) > 0,$$

where $\beta_i = \prod_{j \neq i} (\alpha_i-\alpha_j)$, $1 \leq i \leq n$; we have as in Theorem 15

$$|\mathcal{W}| \leq 2(4 - 3/n)H.$$

We observe that if v lies outside $\mathcal{W}$ and $v(X) < 0$, then $v(X-\alpha_i) = v(X)$ for $1 \leq i \leq n$. Hence if S_i, $i = 1,2,3$, denotes the set of valuations lying outside $\mathcal{W}$ at which $v(X-\alpha_i) > 0$, then the fundamental inequality yields

$$H((X-\alpha_1)(\alpha_2-\alpha_3)/(X-\alpha_2)(\alpha_3-\alpha_1)) \leq |\mathcal{W} \cup S_1 \cup S_2 \cup S_3| + 2g.$$

Now

$$\frac{X-\alpha_2}{X-\alpha_1} - 1 = \frac{\alpha_1-\alpha_2}{X-\alpha_1},$$

and so

$$H((X-\alpha_1)/(\alpha_1-\alpha_2)) = H((X-\alpha_1)/(X-\alpha_2)).$$

Furthermore, we obtain $m|S_i| \leq H(X-\alpha_i)$, $i = 1,2,3$, as before, and thus

$$H(X) \leq (3/m)H(X) + |\mathcal{W}| + (7/n + 3/nm)H + 2g,$$

from which the result follows, since $m \geq 4$.

In the special case $m=n\geq 4$, we obtain

$$H(X,Y) \leq 36H + 8g.$$

Upon substituting $X = Z/T$, $Y = 1/T$, we see that Z, T is a solution of the general Thue equation $\alpha_0 \prod_{i=1}^{n} (Z-\alpha_i T) = 1$. Since $H(X,Y) = H(Z,T)$, it follows that if $F(X,Y)$ is a form of degree at least 4 which factorises over K into distinct linear factors, then any solution in K of $F(X,Y) = 1$ satisfies

$$H(X,Y) \leq 36H + 8g ,$$

where H is the height of F and g is the genus of K/k. Previously Schmidt had obtained the bound $89H + 212g$ when F is a separable form over K of degree at least 5. The condition on the degree may not be weakened further than here, since then $F(X,Y) = 1$ would represent a curve of genus 0 or 1, which may possess solutions of unbounded height. As remarked earlier, the condition on separability may be relaxed in Theorem 16. In fact if F is a binary form, then the method of Theorem 16 yields a bound on the heights of solutions in K of $F(X,Y) = 1$ precisely when this equation has genus 2 or more. For the superelliptic equation, Schmidt's method is actually more powerful at present, for he has shown [37] that effective bounds may be determined, at least in principle, on the solutions of $Y^m = f(X)$, when X, Y and the coefficients of f lie in $k(z)$, and $m = 2$, $n \geq 17$ or $m = 3$, $n \geq 5$, together with $m \geq 4$, $n \geq 3$.

We conclude with a few remarks concerning possible generalisations. If in Lemma 2 we write $\alpha = -\gamma_1/\gamma_3$, $\beta = -\gamma_2/\gamma_3$, so that α and β satisfy $\alpha + \beta = 1$ and have all their poles and zeros within the finite set V, then the conclusion of Lemma 2 is that

$$H(\alpha) \leq \max\{0, |V|+2g-2\}.$$

The inequality thus applies to the one linear equation in two unknowns $\alpha + \beta = 1$. Two directions for generalisation are therefore suggested, the first being to increase the number of equations, and the second to increase the number of unknowns. Either would have important consequences for equations over function fields. For the former, a sufficiently strong bound for the solutions of three simultaneous linear equations in two unknowns would lead to a polynomial bound for the heights of all the solutions in K of any equation of genus 2. For the latter, any effective bounds for the solutions of a single equation in an arbitrary number of unknowns would lead to a complete resolution of the general norm form equation over function fields. Both problems still remain open in the case of number fields, although they have recently been resolved ineffectively by Faltings and Evertse respectively. The former problem is open for function fields also, but I am pleased to be able to conclude by announcing my successful resolution of the latter problem for function fields.

REFERENCES

1 Armitage, J.V. The Thue-Siegel-Roch theorem in characteristic p, J. Algebra 9 (1968), 183-189.

2 Armitage, J.V. An analogue of a problem of Littlewood, Mathematika 16 (1969), 101-105; Corrigendum and Addendum, ibid. 17 (1970), 173-178.

3 Baker, A. Linear forms in the logarithms of algebraic numbers, I, II, III, IV, Mathematika 13 (1966), 204-216; 14 (1967), 102-107; 14 (1967), 220-228; 15 (1968), 204-216.

4 Baker, A. Contributions to the theory of Diophantine equations: I On the representation of integers by binary forms; II The Diophantine equation $y^2 = x^3 + k$, Philos. Trans. Roy. Soc. London Ser. A 263 (1968), 173-208.

5 Baker, A. The Diophantine equation $y^2 = ax^3 + bx^2 + cx + d$, J. London Math. Soc. 43 (1968), 1-9.

6 Baker, A. Bounds for the solutions of the hyperelliptic equation, Proc. Camb. Philos. Soc. 65 (1969), 439-444.

7 Baker, A. Transcendental number theory, Cambridge University Press, Cambridge, 1975.

8 Baker, A. and Coates, J. Integer points on curves of genus 1, Proc. Camb. Philos. Soc. 67 (1970), 595-602.

9 Coates, J. Construction of rational functions on a curve, Proc. Camb. Philos. Soc. 68 (1970), 105-123.

10 Davenport, H. On $f^3(t)-g^2(t)$, Norske Vid. Selsk. Forh. (Trondheim) 38 (1965), 86-87.

11 Eichler, M. Introduction to the theory of algebraic numbers and functions, Academic Press, New York, 1966.

12 Fröhlich, A. and Shepherdson, J.C. Effective procedures in field theory, Philos. Trans. Roy. Soc. London Ser. A 248 (1956), 407-432.

13 Gill, B.P. An analogue for algebraic functions of the Thue-Siegel theorem, Ann. of Math. Ser. 2 31 (1930), 207-218.

14 Kolchin, E.R. Rational approximation to solutions of algebraic differential equations, Proc. Amer. Math. Soc. 10 (1959), 238-244.

15 Landau, E. and Ostrowski, A. On the Diophantine equation $ay^2 + by + c = dx^n$, Proc. London Math. Soc. Ser. 2 19 (1920), 276-280.

16 Lang, S. Abelian varieties, Interscience, New York, 1959.

17 Lang, S. Algebraic number theory, Addison-Wesley (1970).

18 Lefschetz, S. Algebraic geometry, Princeton University Press, Princeton, New Jersey, 1953.

19 Liouville, J. Sur des classes très-étendues de quantités dont la valeur n'est ni algébrique, ni même reductible à des irrationnelles algébriques, Comptes rendus de l'Académie des Sciences (Paris) 18 (1844), 884-885, 910-911, J. Math. pures appl. 16 (1851), 133-142.

20 Mahler, K. On the theorem of Liouville in fields of positive characteristic, Canadian J. Math. 1 (1949), 397-400.

21 Mason, R.C. On Thue's equation over function fields, J. London Math. Soc. Ser. 2 24 (1981), 414-426.

22 Mason, R.C. The hyperelliptic equation over function fields, Proc. Camb. Philos. Soc. 93 (1983), 219-230.

23 Mason, R.C. Equations over function fields, Approximations Diophantiennes et nombres transcendants, Birkhäuser, Progress in Mathematics Vol. 31, 143-149.

24 Maunder, C.R.F. Algebraic topology, Van Nostrand, London, 1970.

25 Mordell, L.J. The Diophantine equation $y^2 - k = x^3$, Proc. London Math. Soc. Ser. 2 13 (1913), 60-80.

26 Mordell, L.J. Indeterminate equations of the third and fourth degrees, Quart. J. Pure and Appl. Math. 45 (1914), 170-186.

27 Mordell, L.J. Note on the integer solutions of the equation $Ey^2 = Ax^3 + Bx^2 + Cx + D$, Mess. Math. 51 (1922), 169-171.

28 Mordell, L.J. On the rational solutions of the indeterminate equations of the third and fourth degrees, Proc. Camb. Philos. Soc. 21 (1922), 179-192.

29 Mordell, L.J. On the integer solutions of the equation $ey^2 = ax^3 + bx^2 + cx + d$, Proc. London Math. Soc. Ser. 2 21 (1923), 415-419.

30 Mumford, D. Abelian varieties, Oxford University Press, India, 1970.

31 Osgood, C.F. An effective lower bound on the "Diophantine approximation" of algebraic functions by rational functions, Mathematika 20 (1973), 4-15.

32 Osgood, C.F. Effective bounds on the "Diophantine approximation" of algebraic functions over fields of arbitrary characteristic and applications to differential equations, Indag. Math. 37 (1975) no. 2, 105-119; Errata, ibid. 37 (1975) no. 5, 401.

33 Roth, K.F. Rational approximations to algebraic numbers, Mathematika 2 (1955), 1-20; Corrigendum, ibid. 2 (1955), 668.

34 Samuel, P. Lectures on old and new results on algebraic curves, Bombay, 1966.

35 Schmidt, W.M. On Osgood's effective Thue theorem for algebraic functions, Comm. Pure Appl. Math. 29 (1976), 759-773.

36 Schmidt, W.M. Thue's equation over function fields, J. Austral. Math. Soc. Ser. A 25 (1978), 385-422.

37 Schmidt, W.M. Polynomial solutions of $F(x,y) = z^n$, Queen's Papers in Pure Appl. Math., 54 (1980), 33-65.

38 Siegel, C.L. Approximation algebraischer Zahlen, Math. Zeitschrift, 10 (1921), 173-213.

39 Siegel, C.L. The integer solutions of the equation $y^2 = ax^n + bx^{n-1} + \ldots + k$, J. London Math. Soc. 1 (1926), 66-68 (under the pseudonym X).

40 Siegel, C.L. Über diophantische Gleichungen, Abh. Preuss. Akad. Wiss. (1929) no. 1, 41-70.

41 Silverman, J.H. The Catalan equation over function fields, Trans. A.M.S., 273 (1982), 201-205.

42 Thue, A. Über Annäherungswerte algebraischer Zahlen, J. Reine Angew. Math. 135 (1909), 284-305.

43 Thue, A. Über die Unlösbarkeit der Gleichung $ax^2 + bx + c = dy^n$ in grossen ganzen Zahlen x und y, Arch. Math. Naturvid, Kristania Ser. B 34 (1917), no. 16.

44 Uchiyama, S. On the Thue-Siegel-Roth theorem, I, II, III, Proc. Japan Acad. 35 (1959), 413-416; 35 (1959), 525-529; 36 (1960), 1-2.

45 Uchiyama, S. Rational approximations to algebraic functions, J. Fac. Sci. Hokkaido Univ. Ser. 1 15 (1961), 173-192.

46 van der Waerden, B.L. Moderne Algebra Vol. II, Springer, Berlin, 1931.

47 Weil, A. L'arithmétique sur les courbes algébriques, Acta Math. 52 (1928), 281-315.

Printed in the United States
By Bookmasters